Christian Huber

Throughput Analysis of Manual Order Picking Systems with Congestion Consideration

AF302743

Wissenschaftliche Berichte des
Institutes für Fördertechnik und Logistiksysteme
des Karlsruher Instituts für Technologie
Band 76

Throughput Analysis of Manual Order Picking Systems with Congestion Consideration

by

Christian Huber

Dissertation, Karlsruher Institut für Technologie
Fakultät für Maschinenbau, 2011
Referent: Prof. Dr.-Ing. K. Furmans
Korreferent: Prof. Dr. K. J. Roodbergen

Impressum

Karlsruher Institut für Technologie (KIT)
KIT Scientific Publishing
Straße am Forum 2
D-76131 Karlsruhe
www.ksp.kit.edu

KIT – Universität des Landes Baden-Württemberg und nationales
Forschungszentrum in der Helmholtz-Gemeinschaft

KIT Scientific Publishing 2011
Print on Demand

ISSN: 0171-2772
ISBN: 978-3-86644-759-2

Throughput Analysis of Manual Order Picking Systems with Congestion Consideration

Zur Erlangung des akademischen Grades eines

Doktors der Ingenieurwissenschaften

der Fakultät für Maschinenbau
des Karlsruher Instituts für Technologie (KIT)
genehmigte

Dissertation

von

Dipl.-Wi.-Ing. Christian Huber

Tag der mündlichen Prüfung: 27. Juli 2011
Hauptreferent: Prof. Dr.-Ing. K. Furmans
Korreferent: Prof. Dr. K.J. Roodbergen

Vorwort

Die vorliegende Arbeit entstand während meiner Tätigkeit als wissenschaftlicher Mitarbeiter am Institut für Fördertechnik und Logistiksysteme des Karlsruher Instituts für Technologie. Ich möchte mich an dieser Stelle bei allen Personen bedanken, die zum Gelingen dieser Arbeit beigetragen haben.

Herrn Prof. Dr.-Ing. Kai Furmans, Leiter des Instituts für Fördertechnik und Logistiksysteme, gilt mein besonderer Dank für die Übernahme des Hauptreferats sowie für die Unterstützung meiner Tätigkeit als wissenschaftlicher Mitarbeiter. Die Möglichkeit zum selbstständigen Arbeiten, die zielführenden Diskussionen sowie die richtigen Fragen haben maßgeblich dazu beigetragen, meine Promotion erfolgreich abzuschließen.

Herrn Prof. Dr. Kees Jan Roodbergen danke ich für die Übernahme des Korreferats sowie die interessanten Diskussionen zu manuellen Kommissioniersystemen. Herrn Prof. Dr. rer. nat. Alexander Wanner danke ich für die Übernahme des Prüfungsvorsitzes.

Mein herzlicher Dank gilt allen ehemaligen Kollegen für die sehr angenehme Arbeitsatmosphäre, Ihre Unterstützung sowie die zahlreichen Anregungen. Hervorheben möchte ich Jens Wisser, der maßgeblich für die Beantragung des Forschungsprojektes, welches dieser Arbeit zugrunde liegt, verantwortlich war und mir fortlaufend wertvolle Hinweise gegeben hat. Besonders danke ich auch Melanie Schwab und Judith Stoll für die Korrektur des Manuskriptes. Mein Dank gilt auch Eda Özden für die hilfreichen Diskussionen über geschlossene Bediensystemnetzwerke. Ebenso danke ich den studentischen Hilfskräften, die meine Forschungsarbeit begleitet haben. Im Speziellen Martin Epp, der mir bei der Umsetzung der Ideen zur Berechnung der Netzwerkparameter eine große Hilfe war. Florian Denz, Steffen Hedrich und Murtaza Bootwala danke ich für die fruchtbaren Diskussionen zu Korrekturfaktoren in geschlossenen Bediensystemnetzwerken.

Mein persönlicher Dank gilt meiner Familie und insbesondere meinen Eltern, die mir den eingeschlagenen Bildungsweg als Grundstein dieser Arbeit ermöglicht haben. Sie, meine Schwiegereltern und meine Schwägerin haben mich stets unterstützt, Interesse am Fortschritt meiner Arbeit gezeigt und an mich geglaubt. Mein größter Dank gilt meiner Frau Corinna und meinem Sohn Jannik für ihre Liebe, Unterstützung und für ihr Verständnis. Sie haben mir die notwendige Kraft und Motivation zum Gelingen dieser Arbeit gegeben und zeigen mir jeden Tag aufs Neue, was die wirklich wichtigen Dinge im Leben sind.

Waldbronn, September 2011 Christian Huber

Kurzfassung

Christian Huber

Durchsatzanalyse manueller Kommissioniersysteme unter Berücksichtigung von Blockiereffekten

Kommissionierung ist der wichtigste Prozess in Distributionszentren. Typischerweise sind Kommissioniersysteme manueller Art, da Arbeitskräfte zu den Kommissionierplätzen laufen oder fahren, um dort Artikel für einen Kundenauftrag zu entnehmen. Im Rahmen der Planung solcher Systeme stellt die Gangbreite einen relevanten Entscheidungsparameter dar. Durch enge Gänge können Flächenkosten und Wegstrecken reduziert werden. Jedoch kann es zu Blockiervorgängen kommen, da Kommissionierer sich gegenseitig nicht überholen können. Dadurch resultierende Wartezeiten haben einen negativen Einfluss auf das Durchsatzverhalten. Der Großteil existierender Forschungsarbeiten behandelt Systeme mit einem Kommissionierer und ermittelt die zurückzulegende Wegstrecke ohne auf Abhängigkeiten zwischen Kommissionerern einzugehen. Nur sehr wenige Veröffentlichungen berücksichtigen Blockiereffekte, haben aber aufgrund getroffener Annahmen einen eingeschränkten Anwendungsbereich.

In der vorliegenden Arbeit werden bedientheoretische Modelle genutzt, um den Durchsatz in manuellen Kommissioniersystemen mit Blockiereffekten zu berechnen. Die Modelle behandeln Systeme, denen eine Durchgangstrategie sowie eine zufällige bzw. klassenbasierte Belegungsstrategie zugrunde liegen. Die Verfahren sind analytischer Natur und ermöglichen daher die Untersuchung zahlreicher Alternativen in vergleichsweise kurzer Zeit.

Im Rahmen der Modellierung werden die Parameter eines manuellen Kommissioniersystems Schritt für Schritt in die Parameter des bedientheoretischen Modells übertragen. Die Modellierung führt dazu, dass einzelne Bedienstationen keine Puffer und die Bedienzeiten eine generelle Verteilung besitzen. Bestehende Approximationsansätze werden erweitert und in einer neuen Methode integriert. Eine sehr gute Ergebnisgüte für den Großteil der durchgeführten Experimente validiert den entwickelten Ansatz.

Die neue Approximationsmethode wird anhand zahlreicher Parameterkonfigurationen angewandt, um die Effeke der Blockierungen zu quantifizieren. Die Ergebnisse zeigen, dass die Durchsatzverluste auch für eine geringe Anzahl an Kommissionierern im System nicht zu unterschätzen sind. Bestehende Handlungsempfehlungen zu Planung und Betrieb von manuellen Kommissioniersystemen sind zum Teil neu zu definieren, wenn Blockiereffekte auftreten. In Systemen mit wenigen langen Gängen treten Blockierungen häufiger auf als in Systemen mit vielen kurzen Gängen. In Bezug auf die Belegungsstrategie wird gezeigt, dass eine klassenbasierte Belegung nicht zwangsläufig besser abschneidet als eine zufällige. So kann es für bestimmte Szenarien empfehlenswert sein, häufig nachgefragte Artikel gleichmäßig auf das System zu verteilen, anstatt sie in bestimmten Zonen zusammenzufassen.

Abstract

Christian Huber

Throughput Analysis of Manual Order Picking Systems with Congestion Consideration

Order picking is the most important warehouse process. It is typically organized in a manual way, meaning that humans walk or drive to a location, collecting items for customer orders. When designing manual order picking systems, the width of an aisle is one parameter of interest. Narrow aisles will initially reduce surface costs and travel distance. However, such configurations might suffer from congestion as pickers cannot pass each other, resulting in waiting times which negatively influence throughput performance. The majority of existing literature deals with single-picker operations, mostly focusing solely on travel distance and thus not regarding interdependencies between pickers. A few publications have incorporated congestion but they have very strict assumptions, resulting in a limited scope of application.

In this work, we use queueing models to calculate throughput of manual order picking systems with congestion consideration. We analyze systems with traversal routing and random as well as class-based storage policy. Being of analytical nature, the models are able to estimate throughput for many alternative designs in a relatively short amount of time.

We perform a step-by-step transformation of order picking systems' parameters into parameters of the queueing network. This results in some distinct queueing model characteristics, namely non-existing buffers and generally distributed service times. Existing approximation approaches are extended and integrated to obtain a new method for this type of network. The accuracy of the new approach is analyzed, showing a small error for the overwhelming majority of experiments.

The approach is applied to numerous order picking system parameter configurations to quantify the effects of congestion. The results suggest that a considerable percentage of throughput is lost, even for a small number of pickers in the system. Furthermore, existing guidelines on how to design and operate manual order picking systems are at least partly not valid if congestion occurs. In terms of layout design, we conclude that systems with few long aisles suffer more from congestion than systems with many short aisles. Furthermore, we show that in terms of throughput, class-based storage does not necessarily outperform random storage. It might therefore be advisable to distribute frequently picked items evenly over the whole system instead of clustering them in distinct zones.

Contents

1 Introduction

When managers and experts from traditional industrial countries learned that mass production was not the only and most efficient organizational strategy to fulfill customer demand, the so-called Lean Management became popular. Being associated with terms such as *waste elimination* and *minimum inventory*, some people even started to dream about warehouse-free supply chains. Warehouses with large inventories, waiting for customers who might not even exist, were identified quite simply as waste, which in theory should be eliminated. However, people pursuing this vision were never able to answer questions like how a production facility could carry on operating if important suppliers were located overseas and a transportation problem emerged. Or whether a customer in Germany would be willing to wait several days for a car spare part to be delivered from an Eastern European production site.

Today, the importance of warehouses for efficient, robust and quick supply chain processes has mostly been realized. Frazelle (2002, p. 1) states that warehouses will always play a major role in connecting companies to consumers and according to Sheth (1995, p. 153), the cost of keeping stock is preferred to the unavailability of material. Even Taiichi Ohno (1988, p. 128) - founder of the famous Toyota Production System and inspirator for many inventory reduction efforts - concludes that a certain amount of inventory is needed to enable a running process. Consequently the focus should not be on eliminating warehouses in general but to reduce the waste within its processes.

Order picking is the most important warehouse process. Gudehus (2005, p. 685) defines it as the process of retrieving items from an assortment of goods according to individual customer orders. The main reason for its importance is the fact that on average 45-55% of total warehousing costs can be attributed to order picking (de Koster et al. 2007, p. 481,

Warehouse Excellence Study 2011). Additionally, in many industries order patterns are changing from a few large orders to many small orders (van den Berg 1999, p. 751), thus increasing the importance for flexible and efficient order picking operations. To concentrate on the efficiency of such operations when trying to continuously improve the warehouse performance is therefore advisable.

Order picking systems are typically operated such that a picker walks or drives to the goods (picker-to-part system) and manually retrieves them from a storage location. Because humans are involved in every single step of this operation, we will use the term *manual order picking system*. De Koster et al. (2007, p. 485) number the portion of manual systems used in Western European order picking operations at 80%. Data from the Warehouse Excellence Study (2011) support these statements as of 220 order picking tasks surveyed approx. 83% use a picker-to-part strategy. Despite many efforts towards more automated warehouses the order picking process as a whole remains a very labor-intensive operation.

A major performance indicator in manual order picking systems is the throughput. It is either measured by counting the completed orders or the number of completed order lines per period of time. Both figures can be used to judge whether the order picking system is able to perform according to a certain customer demand. Being a labor-intensive operation, manual order picking usually involves not only one single picker but multiple pickers working simultaneously in the system. Increasing the number of pickers seems to be an intuitive way to increase throughput and reduce order cycle time respectively. However, it might lead to congestion which negatively affects throughput. In this thesis we present analytical models capable of calculating throughput in an order picking system with congestion consideration.

1.1 Problem Description

As in many processes involving multiple participants, the pickers have to *share* the system and thus do not work independently of each other. Initially, these interactions prevent an undisturbed order picking process which we would experience if pickers worked alone. The interdependencies are apparent when pickers block each other at various locations of the order picking system. Such a blocking situation results in congestion, which ultimately reduces the systems' throughput.

Besides the number of workers, the level of congestion heavily depends on the width of the aisles. It is especially distinct in order picking systems with narrow aisles that do not allow for passing of two or more pickers. Thus one or several pickers have to wait for another picker even if they do not need to access the occupied storage location. Some operational strategies, e.g. a specific zone for high-demand items, might also worsen the situation because then more pickers are likely to stay within a small part of the system.

As a consequence, we might wonder why not all order picking systems are built with wide aisles, thus only suffering from minor waiting times while preventing heavy congestion. Even though this seems logical at first sight, we should bear in mind the costs attributed to space. Each square meter has to be either built or rented and constantly generates surface costs, e.g. for insurance, air-conditioning or cleaning. There is a trade-off between the costs of congestion and the costs of surface: if we build wide-aisle systems, we will not suffer much congestion but will have increased surface costs. If we operate a narrow-aisle system, we will have more congestion but save surface costs. The effects of congestion can presumably be quite high in layouts with narrow aisles. Because of this, throughput in such systems is of special interest when evaluating different design alternatives within the rough planning phase. Simulation models are a possibility to estimate throughput, however they are quite time consuming in both model building and experimenting. It is thus desirable to have analytical models capable of quickly calculating throughput for numerous alternative designs of narrow-aisle systems.

The overwhelming majority of existing scientific literature on manual order picking systems has dealt with single-picker operations focusing on travel distance without any consideration of congestion phenomena. Only very few approaches have incorporated the effects of congestion into the quantitative analysis of order picking systems. Besides some simulation studies, just a few analytical approaches exist and they are rather restricted in terms of input parameters and have tight assumptions.

For instance they do not allow to choose arbitrary picking times or assume that a picker visits all picking locations on each tour. Consequently it is rather difficult to use these models for the evaluation and comparison of different design alternatives.

The purpose of this thesis is to develop an analytical model of manual order picking systems with congestion consideration by means of continuous time queueing theory. Various input parameters of manual order picking systems will be transformed into the parameters of the queueing model. An approximation scheme is presented to calculate throughput of such systems. We apply the models to a large set of input parameters, enabling us to gain some insights on the behavior of systems with congestion. Thus, we can show to what extent the waiting times caused by congestion influence the throughput performance.

1.2 Organization of the Thesis

An overview on manual order picking systems, focusing on the main design decisions and operational policies such as layout, storage assignment and routing will be presented in chapter 2. The aspects addressed in this chapter are of interest as they will serve as reference points for the development of the analytical model.

Chapter 3 will give an extensive literature review on models that calculate performance measures for manual order picking systems. At first, a review of models without congestion consideration focusing on travel distance for single-picker operations is given. Subsequently the focus will be on existing models with congestion consideration. In particular, we will identify the weaknesses of these approaches and argue why they offer only limited application possibilities for typical questions of a rough planning phase. The chapter will close with some requirements for the analytical model.

The queueing model of manual order picking systems with congestion consideration will be introduced in chapter 4. We will briefly review the essential basics of queueing theory and explain our motivation to use this methodology. Step by step, the resources and parameters of the manual order picking system are transferred into the parameters of the queueing model. First, recurring elements of the order picking system, e.g. a rack and the corresponding space in front of it, are transformed into recurring modules of a closed queueing network. Subsequently, we calculate routing probabilities for a traversal routing policy assuming either a random storage as well as a class-based storage policy. The chapter also describes how order picking times can be modeled as service times of the queueing model. Finally, the parameter transformation is validated.

Chapter 5 will present an approximation approach which calculates throughput in closed queueing networks with congestion. Existing solution schemes cannot be directly applied because of the assumptions resulting from the modeling in the previous chapter. We first focus on one specific existing algorithm - Rall's integrated approach - and analyze its accuracy. These insights will help us to develop a new integrated approach, which combines existing methods with empirical correction factors specifically derived for manual order picking systems. Closed-form expressions will relate these factors to important input parameters of the order picking system. Ultimately, the factors include the effects of blocking situations and thus congestion involving several pickers.

The new integrated approach will be validated in chapter 6. For this purpose, a large set of input parameter configurations is analyzed with the presented queueing model. The analytical results are compared with results from simulation. A detailed analysis on how each input parameter influences the accuracy is also conducted. Furthermore, the chapter includes an application of our new approximation scheme. We will determine to what extent congestion will reduce throughput and which parameters have the biggest effects. This will give us some insights on how narrow-aisle systems behave. We will also discuss if well-known operating strategies, which were derived from single-picker models, are still valid in narrow-aisle systems with multiple pickers.

The work closes by recapitulating the main results and findings. Furthermore, we will discuss the limitations of the presented approach and give an outlook on possible future research.

2 Manual Order Picking Systems

To enable a better understanding of the throughput models presented in this thesis, we will first give an introduction on manual order picking systems. We will describe the key characteristics of such systems (2.1) and discuss the basic design (2.2) and main operational decisions (2.3).

2.1 Characteristics of Manual Order Picking Systems

Order picking can be distinguished into three main classes: picker-to-part systems, part-to-picker systems and automated item dispensing systems. In picker-to-part systems the order picker moves (by walking or driving) to the different locations and directly picks the items from shelves or racks respectively. In part-to-picker systems the order picker remains at a certain location while items are picked by an automated storage and retrieval machine (AS/RS) and then transported to the order picker. Examples of such systems include horizontal or vertical carousel storage, as well as automated storage and retrieval systems for large load carriers (e.g. pallets) and miniload (end-of-aisle) systems for small load carriers (e.g. containers or bins) (Roodbergen 2001, p. 9). In automated item dispensing systems items of more or less uniform size can be automatically dispensed on a conveyor. Note that the term automatic might be misleading, as the replenishment of such systems is mostly performed by humans (Frazelle 2002, p. 144).

A brief look at the literature published on order picking reveals that most papers are actually focusing on part-to-picker systems (Le Duc 2005, p. 28, Wäscher 2004, p. 324). This is in stark contrast to observations of real-life systems. Order picking is performed manually, i.e. in picker-to-part systems, in most cases (Le-Duc 2005, p. 4, Roodbergen 2001, p. 124, Hsieh and Tsai 2006, p. 626).

There are several advantages of manual systems which justify their popularity. First, the investment in hard- and software is relatively small. Secondly, manual systems yield a high flexibility in terms of variable throughput requirements as well as a fluctuating assortment of goods. Additionally, the flexibility of such systems often enables shorter order cycle times and increased on-time-deliveries (Gudehus 2005, p. 696). Disadvantages of manual systems include a relatively low space utilization and big portions of walking time, which Bartholdi and Hackman (2007, p. 151) consider as waste. Part-to-picker systems have the advantage of virtually non-existent travel time (for workers) as well as high throughput performance. On the other hand such systems usually have high investment costs, relatively large order cycle times and a limited flexibility concerning highly variable throughput requirements (Gudehus 2005, p. 699). The issue of investment costs becomes increasingly important as many warehousing services are outscourced to third-party logistics providers. Duration of contracts, typically less than three years, often prevent large investments.

Certainly part-to-picker systems would be used in larger numbers if they could be designed such that they offered higher flexibility at reasonable costs. Current research activities on future order picking systems focus on small-scale autonomous redundant elements (Furmans et al. 2009, Mayer 2009). These new technologies could possibly overcome the barrier between existing picker-to-part and part-to-picker systems.

In the following we will provide more insights on the structure and operations of picker-to-part, i.e. manual order picking systems. For a review on part-to-picker systems see e.g. Cormier and Gunn (1992), van den Berg (1999), Rouwenhorst et al. (2000) as well as Gu et al. (2007, 2010).

2.2 Design of Manual Order Picking Systems

2.2.1 Structural Decisions

Structural decisions typically have long-term character, spanning over a period of approximately five years (Rouwenhorst et al. 2000, p. 521). In this stage a planner has to determine how the system should be integrated with other warehouse processes and how the system itself should be designed. Both decisions are strongly influenced by customer requirements, e.g. assortment of goods, orders, throughput and order cycle times (Gudehus 2008, p. 669).

A typical decision in this phase focuses on dividing an order picking system into subsystems, namely a forward and a reserve area. The reserve area usually stores items on large load carriers like pallets, frequently replenishing fractions of the stock to a forward area, which has limited storage space. This will lead to additional handling effort for replenishment cycles, but at the same time will reduce travel distances in the compact forward area (van den Berg and Zijm 1999, p. 7, Gu et al. 2007, p. 6). For literature on the forward-reserve problem, the reader is referred to van den Berg et al. (1998) as well as Frazelle and Hackman (1994).

The sizing decision insures that systems offer sufficient space for actual and future assortments of goods. The allocation of space to warehouse processes undoubtedly is one of the first decisions in warehouse design. It is therefore crucial to allocate the right amount of space as later changes might be impossible or very expensive. Ghiani et al. (2004, p. 166) present methods to derive the number of required storage locations considering peak inventory levels. Another approach by Arnold and Furmans (2009, p. 176) estimates the number of storage locations such that the system can accommodate incoming stock at a given security level. For further reading on sizing we refer to Roodbergen (2001, p. 12).

The choice which storage or rack system to use is typically influenced by the patterns of customer orders as well as product characteristics like dimensions and weight. If orders can be served by full pallets, which did not undergo separation, block stacking, where pallets are stacked on top and/or side by side of each other might be a feasible solution. Berry (1968) presents some general thoughts about block stacking layouts, Stadler (1996) discusses deep lane storage systems and Goetschalckx and Ratliff (1991) focus on optimal lane depths. The latter as well as Ashayeri and Gelders (1985) also present good literature reviews.

The models presented in this thesis are based on systems where items are stored on shelves or racks which are arranged in aisles. We can distinguish between low-level and high-level systems (Le-Duc 2005, p. 5).

In low-level systems, items are stored and picked from bin shelving, modular storage drawers or gravity flow racks. The order picker remains on the floor as systems have a height of less than 1.8m (Roodbergen 2001, p. 9). Bin shelving is comparatively cheap, easily reconfigurable and requires almost no maintenance (Frazelle 2002, p. 128). It allows for the storage of a wide spectrum of goods, e.g. non-palletized items, small items, paperboard containers, bins and even bulk items (ten Hompel et al. 2007, p. 64). Space utilization is the major disadvantage of bin shelving as space requirements rapidly grow with the assortment of goods and the in-shelf space is rarely used to its full extent. Therefore high surface costs as well as extended travel times and thus labor costs occur (Frazelle 2002, p. 128). In modular storage drawers space utilization is improved as no additional space is needed for reaching into the drawers. However, Frazelle (2002, p. 130) notes they are not as flexible as bin shelving as they are limited to small items. Gravity flow racks are typically used for broken case picking of items with high demand, maintaining a first-in-first-out principle by replenishing from the back of the rack. They provide reasonable investment costs, maintenance and space utilization (ten Hompel et al. 2007, p. 81). In low-level systems, the typical material handling equipment would be a picking cart on which an order picker accumulates the different items of his tour. Alternatively a system can be equipped with automatic conveyor technology to enable tote picking (Frazelle 2002, p. 134). The space utilization in low-level systems can often be increased by implementing one or more mezzanines (Frazelle 2002, p. 131). Time consuming movements between levels can be reduced by placing fast movers on the floor level and slow movers on the mezzanines.

Most high-level systems, e.g. single-deep or double-deep pallet racks, high-level bin shelving, pallet flow racks as well as mobile pallet racks are similar to low-level systems except for their height, which according to Thomas (2008, p. 649) might be as high as 40m. They also differ in the size of the stored loading units, which in most cases is a pallet. Picking from pallets is typically more time consuming and requires different equipment, e.g. counterbalance lift trucks, stackers or order picking trucks. Apart from that, advantages like flexibility or disadvantages like low space utilization can be transferred from low- to high-level systems. Other high-level systems which combine the advantages of rack storage and block stacking, e.g. drive-in and drive-thru racks, will not be further discussed in this thesis. Instead we suggest the following literature: van den Berg (1996, p. 18), Roodbergen (2001, p.8), Frazelle (2002, p. 90ff.), Gudehus (2005, p. 839), ten Hompel et al. (2007, p. 70), Thomas (2008, p. 636) and Schulte (2009, p. 235).

In the future existing concepts of block stacking and aisle-based storage might be merged to produce new storage and picking systems. Gue (2006) describes very high density storage systems and presents an algorithm to fill storage space subject to limited number of interfering items. In (2007) Gue and Kim analyze interactions between density and retrieval times in puzzle-based-storage systems. Current research is dedicated to the concept of virtual aisles, where aisles form and collapse as needed (Gue 2010).

2.2.2 Layout Decisions

The period of time affected by layout decisions can range from a week or month up to a few years, depending on the flexibility of the system's hardware. Layouts in storage and picking shall be designed to simultaneously maximize space utilization and the level of service (Hwang et al. 2004, p. 3874).

According to de Koster et al. (2007, p. 487) the following issues are part of the internal layout design or aisle configuration problem. In manual order picking systems the floor space is typically utilized by the following elements: storage racks, aisles, cross-aisles, depots and material handling technology (if needed).

Several storage locations on top of each other will be referred to as a rack column. A storage rack then consists of several adjacent rack columns, i.e. pick faces. Two opposing storage racks make up an aisle. In manual order picking systems we can distinguish picking aisles, replenishment aisles as well as cross aisles (Gudehus 2005, p. 745). Cross aisles connect adjacent aisles. According to Roodbergen and Vis (2006, p. 801) the total aisle length describes the total length along the pick face and can be derived from the number of necessary storage locations.

Building upon the total aisle length, the aisle configuration problem determines how many aisles should be used, how long and wide they should be and how they should be orientated. For a total aisle length S, an aisle configuration with ν aisles each of length l is feasible if $S \bmod (\nu \cdot l) = 0$. This equation theoretically allows numerous solutions but in practice only few aisle configurations are realizable as the size and shape of the order picking system often is pre-determined by the size and shape of the total warehouse as well as building restrictions.

The overwhelming majority of published work on order picking assumes systems which comply with the following two design rules. First, aisles are straight and parallel. Second, cross aisles are also straight and meet picking aisles at right angles (Gue and Meller 2009, p. 172). Only few papers have proposed new orientation possibilities, namely the Flying-V, fishbone design and chevron aisles of Gue and Meller (2009) as well as the $\wedge$-Shape as discussed in Ivanovic (2007). Some thoughts on non-conventional aisle design were first analyzed by Berry (1968) and especially White (1972).

Some literature exists on guidelines for conventional layout design. Kunder and Gudehus (1975) provide formulas to derive the optimal number of aisles for different routing strategies. Hall (1993, p. 84) found that good warehouse shapes become elongated with increasing number of picks. Caron et al. (2000b, p. 116) (2000a, p. 103) present formulas that relate the optimal number of aisles to parameters like total aisle length, pick list size and demand skewness. Roodbergen (2001, p. 81) formulates a non-linear programming model to determine optimal layouts for given situations. Hwang et al. (2004, p. 3883) show that a good aisle configuration is one in which system width W is approximately one half of system length L. Roodbergen and Vis (2006, p. 809) investigate the interdependencies between different layouts and routing policies and found that optimal layouts are quite sensitive to the routing policy which was used for the optimization.

Typical one block systems with two cross aisles can be extended by additional cross aisles to form multiple block layouts. There is a trade-off between space and travel time needed for the additional cross aisles and travel time savings due to shortcutting through cross aisles. Even though quite frequently used in real-life, only few authors have discussed this subject. Vaughan and Petersen (1999, p. 884) present a heuristic to estimate travel distances in systems with multiple cross aisles, showing that for a wide range of scenarios multiple cross aisles on average yield 20-30% savings on travel distance. Roodbergen and de Koster (2001b) (2001a) analyze systems with three cross aisles and conclude that in most cases the additional aisle results in less traveled distance. The optimization model of Roodbergen and Vis (2006) is extended by Roodbergen et al. (2008) to allow the consideration of multiple blocks.

The final key element of manual order picking systems is the depot. A depot is a work station where order pickers start and end their tours. We will use depot as a synonym for other terms like dock, central location, loading dock, pick-up/drop-off location, pickup and deposit point or input-output-point. At the depot order pickers return finished orders and receive new orders and picking containers. In system design the number and location of depots has to be determined. Some literature exists on the latter. Bassan et al. (1980, p. 319) and Roodbergen and Vis (2006, p. 804) conclude that for random storage (see chapter 2.3) the depot should be located at the very center of the front cross-aisle. However, performance figures like travel distance appear not to be very sensitive to the depot location as average distance savings for middle depots are less than 1%. These savings can be bigger for very small pick lists or other storage location assignment strategies but tend to be less than 4% (Petersen (1997, p. 1108), Petersen and Schmenner (1999, p. 498) and Petersen and Aase (2004, p. 18)).

In manual order picking systems using pick-to-belt technologies there may be no depot at all as empty containers can be picked up at many points in the system and dropped off decentralized onto a conveyor belt (de Koster and van der Poort 1998, p. 477).

2.3 Storage and Order Picking Operations

Operational policies determine how the system should be run. Some policies are typically applied on a yearly or monthly basis while other policies can be adopted on a day-to-day basis. We will give some insights on the following policies: storage location assignment policies, picker routing policies, order batching policies and zoning policies.

2.3.1 Storage Location Assignment Policies

Random Storage

With a random storage policy, the storage location for incoming loading units is selected randomly from all available empty locations (de Koster et al. 2007, p. 488). Thus all storage locations have an equal probability of being visited in a picking tour. This also holds true if an incoming loading unit is assigned to the closest open location and retrievals are performed on a first-in-first-out (FIFO) basis (Tompkins 2003, p. 420). Random storage, in comparison to other policies, efficiently utilize spaces (Arnold and Furmans 2009, p. 189). Being fairly easy and inexpensive to implement random storage is used in many order picking systems (Roodbergen and Vis 2006, p. 800).

Dedicated Storage

In this policy each item has one or more fixed storage locations, which are exclusively reserved for this specific item. Arnold and Furmans (2009, p. 185) show how for each item the locations have to be chosen as to offer sufficient space to store a certain level of inventory for a given confidence level, resulting in the lowest space utilization of all storage policies. However, due to fix locations order pickers can better familiarize with the items, possibly resulting in shorter searching times and better quality (de Koster et al. 2007, p. 489). In many warehouses dedicated storage is based on item characteristics (e.g. weight, hazardousness, temperature requirements) (Tompkins 2003, p. 431).

Full-Turnover-Based Storage

This policy is a special case of the dedicated storage and also known as volume- or frequency-based storage policy. Items are ranked according to a specific criterion and are then assigned to fixed locations, placing best ranked items closest to the depot and so on. A ranking criterion extensively discussed in the literature is the cube-per-order index, or COI. It considers two basic thoughts: the items stored closest to the depot should be those taking up the least space and those being the most popular ones. The COI is defined as the ratio of an item's required storage space to its order frequency (Kallina and Lynn 1976, p. 42). Items with the lowest COI are placed closest to the depot. Under some assumptions, e.g. single- or dual-command cycles, the COI has been proven to be the cost minimizing storage policy (see the papers of Malmborg et al. (1988, p. 4), (1990, p. 95), (1995, p. 467)). In contrast to COI another ranking criterion might solely consider the popularity (volume or frequency) of an item (Le-Duc 2005, p. 13). In Duration-of-Stay (DOS) policy those items with the smallest ratio of lot size and demand, are assigned to locations close to the depot (Goetschalckx and Ratliff 1990, p. 1120, Petersen 1999, p. 1061). As demands and the assortment of goods constantly change, the ranking might have to be renewed quite often and management has to decide whether travel time savings justify constant reshuffling of storage locations (Roodbergen 2001, p. 15).

Class-Based Storage

In this policy items are grouped into classes, which are then ranked according to the pick frequency of the whole class. A typical assortment of goods often is in line with Pareto's principle which, applied to order picking, states that approximately 20% of items stored account for approximately 80% of turnover or picking activities. The class of fast-movers is located closely to the depot. Within classes storage is random (de Koster et al. 2007, p. 489). Class-based storage performs equally good than full-turnover-based storage policies in terms of travel distance - at the same time being less information intensive (Le-Duc 2005, p. 16, Petersen and Aase 2004).

Depending on the routing policy, Petersen and Schmenner (1999, p. 488) describe four strategies to distribute the classes on the aisles: within-aisle, across-aisle, diagonal and perimeter storage. The within-aisle strategy distributes the items of the most frequent class to aisles closest to the depot. The across-aisle strategy will distribute the items of one class evenly across all aisles, locating items of the most frequent class closest to the cross aisle. For diagonal storage the items will be stored in a diagonal pattern with items of the most frequent class located closest to the depot. Perimeter storage is a special case of the across-aisle strategy when using two cross aisles. The items of the less frequent classes will be located in the middle of the aisles. Within-aisle storage tends to have the lowest route length and works well for all pick list sizes (Jarvis and McDowell (1991, p. 97), Petersen and Schmenner (1999, p. 494), Petersen (1999, p. 1064)

Family Grouping and Contact-Based Storage

Potential relations between products, like a high probability of having two items on the same order might lead to family grouping, in which such items are stored close to each other. Clustering techniques might be applied to identify item similarity (Rosenwein (1994), Kiu (1999)). Contact-based storage considers the direct travels between item locations as the clustering criterion. Some literature on this subject can be found in (Wäscher 2004, p. 334).

2.3.2 Picker Routing Policies

For a given set of items that have to be picked, routing policies describe a sequence on how the picker has to visit the items' locations in order to reduce travel time (Roodbergen 2001, p. 19). Using efficient routes might significantly reduce costs as travel time makes up approximately 50% of total order picking time t_{OPT}. Two main types of routing can be distinguished: optimal and heuristic.

Optimal Routing

The problem of finding a shortest route for a given set of locations can actually be transformed into a special case of the classical Traveling Salesman Problem (TSP). Ratliff and Rosenthal (1983) transform single-block warehouses into graphs and subgraphs and present an algorithm to find shortest routes. Extensions of this basic algorithm include single-block warehouses with decentralised depositing by de Koster and van der Poort (1998), two-block warehouses with a central depot by Roodbergen (2001, p. 66, 2001b) and multiple-block warehouses with a central depot by Roodbergen and de Koster (2001a). Goetschalckx and Ratliff (1988) present an algorithm for wide-aisle warehouses were order pickers need to cross from one side of the aisle to another. The advantages of creating optimal routes are often outweighed be the fact that these routes may appear illogical to order pickers, making them deviate from their given routes trying to find own best routes. Furthermore the algorithms are only applicable to specific sets of warehouses, e.g. rectangular and require a possibly complex implementation in the warehouse management system (Le-Duc 2005, p. 21).

Heuristic Routing

Unlike optimal routings, heuristics will (mostly) lead to non-optimal routes but they offer feasible solutions and require virtually no time to be generated (Roodbergen 2001, p. 31). By making use of a specific set of rules, they result in routes which are easy to understand (Petersen 1997, p. 1102).

In the well-known traversal (or transversal or s-shape) routing policy an aisle containing at least one pick has to be traversed entirely, i.e. the order picker enters the aisle at one end and exists at the other end. We can differentiate between traversal policy with aisle skipping and without aisle skipping. In the latter each aisle is traversed, regardless of the picks. For high pick densities, traversal policy will likely be without aisle skipping (Gudehus 2005, p. 758). Note that the picker might turn around within the aisle after the last pick, thus not traversing this particular aisle entirely.

Under a return routing policy the picker enters and exits the aisle from the same end. This policy can be characterized by single aisle visits or multiple aisle visits. In single aisle visits the picker travels to the pick location farthest from the cross aisle and returns picking all necessary items along the way. In multiple aisle visits, the picker exists the aisle after each pick (Gudehus 2005, p. 736).

The midpoint routing policy is a special case of the return policy. The warehouse is split into two halves. Locations in the upper half are visited using a return policy from the upper cross aisle and locations in the lower half are accessed using a return strategy from the lower cross aisle respectively. In order to switch cross aisles the first and last aisles containing picks are traversed entirely (Roodbergen 2001, p. 34).

The largest gap routing policy is an extension of the midpoint policy. For all aisles containing picks, gaps are identified. A gap can be either the distance between the lower cross aisle and the closest pick, between the upper cross aisle and the closest pick and between any two adjacent picks within an aisle. The picker enters an aisle only as far as the largest gap and turns around. The largest gap thus is the part of the aisle not traversed (Petersen (1997, p. 1102) and Roodbergen (2001, p. 34)).

Petersen, see e.g. (1997, p. 1102), introduced the composite routing policy, which combines the traversal and return policies. This strategy considers the distance between two pick locations in adjacent aisles that have the biggest distance to the lower cross aisle. In order to minimize this distance either a traversal or a return policy is used.

The combined routing policy produces similar routes than the composite heuristic. The decision on which strategy (traversal or return) to use in the next aisle is made by using dynamic programming. This enables to take the next aisle into consideration as well (Roodbergen 2001, p. 44).

Other routing heuristics exist in multiple-block warehouses. Vaughan and Petersen (1999, p. 884) present the aisle-by-aisle heuristic which uses dynamic programming to determine the best cross aisle to use for switching aisles. In (2001, p. 38), Roodbergen describes extended traversal and largest-gap strategies and presents an extension to the combined heuristic. Roodbergen and de Koster (2001a, p. 1877) further improve the combined heuristic, resulting in the combined$^+$ routing policy. In this heuristic the aisles in the block closest to the depot are visited from right to left and the farthest block is not necessarily accessed using the leftmost aisle containing picks.

Studies have proven that for some scenarios optimal routes offer large savings compared to any heuristic, see e.g. (de Koster and van der Poort 1998, p. 477). However, it has been shown that for a broad basis of scenarios, optimal routes offer relatively small savings compared to different routing heuristics, typically yielding a 5% reduction of travel time (Petersen 1997, p. 1109, 1999, p. 493). Petersen and Aase (2004, p. 19) state that methods of batching or storage policies mostly lead to more significant improvements compared to savings based on optimal routings.

There are quite many studies comparing the performance of different routing strategies. We will not present further details as the respective results differ depending on the assumptions and the considered policies as well as their interaction with other operational parameters. Instead the reader is referred to the following papers: Kunder and Gudehus 1975, p. B76, Caron, Marchet and Perego 1998, p. 729, Hall 1993, p. 86, Petersen 1997, p. 1107, de Koster and van der Poort 1998, p. 477, Petersen and Schmenner 1999, p. 492, Petersen and Aase 2004, p. 14, Dukic and Oluic 2007, p. 458, Roodbergen 2001, p. 52, Roodbergen and de Koster 2001a, p. 1879, Goetschalckx and Ratliff 1988, Wäscher 2004, p. 342.

2.3.3 Order Batching Policies

When all items on an order picker's pick list originate from a single customer order, the single order picking (or pick-by-order or strict order picking) strategy is in use. Order batching policies consider several customer orders and try to group them into one pick list. On one tour the picker then simultaneously retrieves items for several customer orders (Le-Duc 2005, p. 17). The items may be sorted directly on the picking vehicle (sort-while-pick) using support tools like trays or levels (Tompkins 2003, p. 439). In contrast, pick-and-sort systems (or two-stage-order picking systems) first pick items of several customer orders batchwise. Afterwards the items are assigned to the customer orders in stage two using some kind of sorting system (Gudehus 2005, p. 723).

An essential benefit of batching is reduced travel time as the distance of a tour combining several orders is smaller than the total distance of all single tours (Gibson and Sharp 1992, p. 58). Especially for orders with few order lines batching can realize economies of scale (Le-Duc 2005, p. 17 and Gudehus 2005, p. 733). On the other hand batching always includes some kind of sorting process and thus additional handling (Frazelle 2002, p. 159). Won and Olafsson (2005, p. 1430) observe that large batches might lead to increased order lead times as picking time increases.

Batch building rules determine how batches should be built up. Exemplary algorithms include the proximity of pick locations of different orders (Choe and Sharp (1991), Le-Duc (2005, p. 18)). As the order batching problem is NP-hard, research has focused on efficient heuristics (de Koster et al. 2007, p. 492). These include rules like random or largest number of items, minimum number of additional aisles or distance between orders (de Koster et al. (1999, p. 1483), Gibson and Sharp (1992, p. 60), Ruben and Jacobs (1999, p. 581)). The time window batch building rule (timeout rule) groups orders that arrive during the same time window (Choe and Sharp 1991). Schleyer (2007, p. 31) proposes two other rules. According to the capacity rule a batch is processed if and only if a specific number of orders has been collected. The minimum batch size rule is a special case of the timeout rule. The batch will be processed if and only if a minimum number of orders has arrived at the end of the time window.

There is extensive literature on batching in order picking systems. Gibson and Sharp (1992, p. 67) found that a distance-based seed algorithm yielded the biggest reductions in tour lengths. Rosenwein (1996, p. 660) uses alternate distance metrics and proposes an improved order batching heuristic. In de Koster et al. (1999, p. 1500) the authors compare several batch building rules and state that the choice of batching heuristic should be made after deciding on the routing policy. Ruben and Jacobs (1999, p. 587) show that batches built on orders with similar cross aisle distances yield low picking times and good picking cart utilization. Petersen and Aase (2004, p. 19) conducted experiments on the interdependencies of batching, storage and routing policies in a sort-while-pick system. They conclude that batching has the biggest potential to reduce total picking time, especially when customer orders tend to be small. Tang and Chew (1997, p. 818) and Le-Duc and de Koster (2007, p. 377) use queueing models to estimate picking times and derive an optimum batch size. Good literature reviews on batching policies are given in Rouwenhorst et al. 2000, p. 526, Roodbergen 2001, p. 25, Le-Duc 2005, p. 17, Gu et al. 2007, p. 12 and de Koster et al. 2007, p. 492.

2.3.4 Zoning Policies

A zoning policy divides an order picking system into a certain number of separated areas and pickers are subsequently assigned to these zones (Frazelle 2002, p. 159). In progressive zoning a customer order sequentially runs through the different zones. In parallel zoning (or synchronized zoning) a customer order may be processed simultaneously in multiple zones (de Koster et al. (2007, p. 491), Tompkins (2003, p. 440)). The consolidation process takes place either while the order is running through the system (progressive zoning) or afterwards in a subsequent area (parallel zoning). In wave picking, parallel zoning is combined with batching, resulting in two-stage-order picking systems as described in chapter 2.3.3 (Tompkins 2003, p. 439). Note that the operational character of zoning changes radically if automatic conveying and sorting technology has to be used, implying high investment costs and the need for integration into warehouse management systems.

As the number of order pickers within a zone is limited, few if any congestion problems will arise (Le-Duc 2005, p. 19). Further advantages include the reduced travel times and bigger familiarization with items in a zone if the storage strategy is not random over several zones. In parallel zoning, order lead times can be reduced assuming an efficient consolidation process. Disadvantages may include longer order lead times in synchronized zoning if imbalanced workloads across zones evolve (Gu et al. 2007, p. 6). The additional handling effort for consolidation should be considered carefully in a trade-off between increased picker productivity and increased investment costs (Frazelle 2002, p. 164).

The literature on zoning is rather limited. De Koster (1994) uses queueing models to estimate throughput in pick-to-belt order picking systems. Jane (2000) proposes heuristic methods to assign items to storage locations to achieve balanced workload across zones. Jane and Laih (2005) extend this approach using clustering techniques to assign items. In a simulation study, Petersen (2002) examined different zone shapes. Le-Duc and de Koster (2005a) present a model to determine the optimal number of zones in parallel zoning. Parikh and Meller (2008) develop a cost model to compare batch to zone order picking and found that in terms of overall costs batch picking is cheaper for low system throughput while zone picking seems to be the better choice if throughput is large.

Using Bucket Brigades, an order picking system with varying zone sizes and intersections will emerge. Inspired by production flow lines, bucket brigade picking makes pickers walk and pick the current order until they meet the first picker in the downstream flow. The order is handed over and the picker walks the opposite direction and takes over the order from the first picker in the upstream flow. The system is self-balancing if pickers are arranged such that the slowest picker starts new orders and the fastest picker finishes them. The concept and its advantages like flexibility and simplicity are described e.g. in Bartholdi and Eisenstein (1996), Bartholdi et al. (2001) or Bartholdi and Hackman (2007, p. 137ff). However, newer studies claim that more discussion is needed to ascertain the gain of bucket brigades in order picking systems (Koo 2009, p. 773).

2.3.5 Workforce Policy

The workforce policy considers the choice on the number of pickers that will work in the system. It also defines the necessary level of qualification.

This important decision is a key driver for achieving throughput goals on the one hand but will ultimately drive operating costs on the other hand. The number of pickers is of special interest in this thesis as it is also the main driver of congestion.

2.4 Chapter Conclusion

We have reviewed the basics of manual order picking systems, namely structural decisions, layout decisions and operational decisions. The decisions within one level partly have a strong interdependency, e.g. on the operational level one storage policy might work very well with a certain routing policy but might not work well with another routing policy. Furthermore, there are interdependencies between decisions on different levels, e.g. the choice of layout has a big influence on the performance of different storage and routing strategies.

When planning manual order picking systems, it is strongly advisable to consider these interdependencies by analyzing the effects simultaneously. Unfortunately very few literature has dealt with this issue, as authors often seek to analyze individual problems in one level, mostly concerning only one particular aspect of that level.

3 Throughput Analysis of Manual Order Picking Systems: a Literature Review

In this chapter we will give a review of models estimating throughput of manual order picking systems. We will first focus on models that do not incorporate congestion in chapter 3.1. A summary of throughput models with congestion consideration will follow in chapter 3.2. A critical evaluation of the existing approaches will result in some open research questions which will help us define the scope of our model.

3.1 Models without Congestion Consideration

We refer to throughput estimation models in this subchapter as *single-picker-models* because the approaches treat the system as if only one picker was working, thus not considering any interdependencies between multiple pickers. The models basically estimate the total distance of a picking tour. The total order picking time t_{OPT} is derived by considering the picker's velocity and times for setup, search and the actual picking of items. Total throughput for multiple-picker operations is then obtained by applying Little's Law (1961), which states that

$$\lambda = \frac{K}{t_{OPT}}$$

where λ represents the order (or order lines or item) throughput and K is the number of pickers.

For the sake of simplicity, the layouts used in this chapter are summarized in figure 3.1 and will be solely identified by the respective letters.

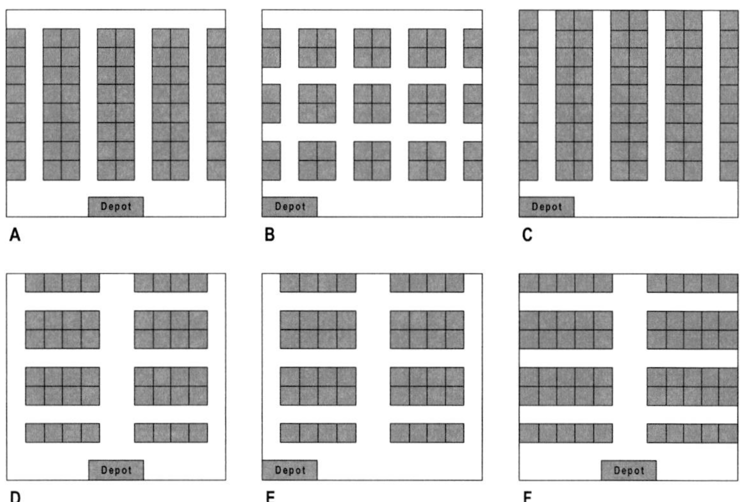

Figure 3.1: Layouts used in travel-time models

3.1.1 Throughput Analysis with Optimal Routing

Some papers have extended the fundamental TSP-approach of Ratliff and Rosenthal (1983), namely de Koster and van der Poort (1998) for decentral depositing, Roodbergen (2001, 2001b) for two-block warehouses as well as Roodbergen and de Koster (2001a) for multiple-block warehouses. Goetschalckx and Ratliff (1988) concentrate on minimal distance paths in wide aisles. These methods all transform order picking systems of layout type A into graphs. The edges are weighted with the actual distance or time between two adjacent locations in the warehouse. Hall (1993) gives a lower bound for the within-aisle travel distance of optimal routes. De Koster et al. (1998) observe that the number of times an aisle has to be entered is minimal for the traversal routing policy. Consequently the time needed to leave and enter aisles for an optimal route is at least as large as for a traversal policy. The authors extend the lower bound of Hall (1993) and give a conservative estimate on total travel time for optimal routes. Vaughan and Petersen (1999) develop an algorithm to find shortest paths in a warehouse with multiple cross aisles (layout type B) by considering distances between cross aisles from which a picking aisle is entered and exited.

3.1.2 Throughput Analysis with Heuristic Routing

Early Work on Unit-Load Warehouses

For the case of unit-load order picking the early works of Francis (1967) and Berry (1968) assume a single command cycle and estimate travel distance in rectangular warehouses of layout type C. Bassan et al. (1980) extends these works to layout types A and D. They also consider different depot locations and multiple zones with one depot each.

Basic Travel-Time Model of Kunder and Gudehus

In the following we concentrate on manual order picking systems with multiple picks per picking tour. The first paper on this subject was published by Kunder and Gudehus (1975). They consider traversal routing policy with and without aisle skipping as well as a return policy with single or multiple aisle visits. Storage location assignment is strictly random and layout type A is used. For a given number of order lines the authors use combinatorial analysis to derive the expected number of aisles to be visited on one tour. This number is then multiplied with the respective within-aisle and partial across-aisle distances to calculate the expected travel distance. Many ideas and procedures of the 1975 paper were adopted, improved and extended by subsequent research.

Travel-Time Models Extending the Work of Kunder and Gudehus

Hall (1993) provided the first extension, also using layout type A and random storage. After inserting a warehouse shape factor into the formulas for traversal strategy, Hall transfers the logic of combinatorial analysis to midpoint and largest gap routing policies. Additionally he shortly discusses travel distances in aisles with non-negligible width where traveling the aisle twice might yield advantages compared to constantly alternating between two opposing storage racks.

The work of Schulte (1996) also builds on the approach of Kunder and Gudehus and extends it by several points. First, it allows two order lines to be on the same rack column, i.e. there might be no travel between two particular picks. Several layout types (A, C, D, F) as well as different depot locations are considered. Schulte assumes random storage and calculates expected within-aisle and across-aisle travel time for traversal and return routing policies. In addition to these extensions, Schulte also corrects some simplifying assumptions of existing research to get more general results. In particular he shows that the formula to calculate the number of aisles to be visited, which was introduced by Kunder and Gudehus (1975, p. B59), is correct for cases where the number of picks is bigger than the number of aisles. It is not valid for cases where the number of picks is smaller than the number of aisles (see Schulte (1996, p. 71ff) for an example in which the error is as high as 25%). Subsequently Schulte derives a new general formula to calculate the expected number of aisles to be visited. Presumably due to the language barrier, Schulte's work is not well-known in the worldwide research community. For some scenarios this might be regrettable as many papers build on the approach by Kunder and Gudehus. On the other hand though, the extent of the error seems to be limited as some models using the approximation while improving other elements of the calculation still report accurate results (see e.g. Roodbergen and Vis (2006, p. 806)).

For traversal routing policy the works of Kunder and Gudehus (1975) and Hall (1993) were further improved by de Koster et al. (1998, p. 393), Roodbergen (2001, p. 82) and Roodbergen and Vis (2006). First, the authors improve the accuracy of within-aisle travel. Kunder, Gudehus and Hall assumed that the last aisle is always traversed completely and both papers absorbed the additional travel by applying a correction factor. In contrast, de Koster et al. (1998) and Roodbergen (2001) both assume that the picker turns immediately after the last pick and adjust the correction factor accordingly. In addition Roodbergen (2001, p. 85) presents an entirely new procedure to estimate travel in cross aisles for any depot location. In contrast to Kunder, Gudehus and Hall, who assumed that the depot is located between the leftmost and rightmost aisles with picks, the author allows for cases where all picks are on only one side of the depot. For the case of decentralized depositing de Koster et al. (1998, p. 390) modify the original formulas such that the expected travel time from the exit of the last aisle of the current order to the entrance of the first aisle of the next order is calculated. An extension of Hall's (1993) travel time estimation for largest gap routing policies was presented by Roodbergen and Vis (2006). They incorporate the fact that the first and last aisles always have to be traversed entirely.

Travel-Time Models for Non-Random Storage Policies

Jarvis and McDowell (1991) developed a method to calculate travel times for traversal routing policy, layout type A and class-based storage. Like Kunder and Gudehus they also use the probability to visit an aisle and distinguish between expected within-aisle and across-aisle travel. The resulting expressions are then used to find product allocations that minimize average order picking time.

The consideration of full-turnover-based storage policies using ABC curves of any skewness was introduced by Caron et al. (1998). They use an analytical function to describe the turnover activity in dependance of the ABC curve shape factor. In their example, the ABC curve is COI-based but ten Hompel and Hömburg (2008, p. 395) note that the function can represent any distribution function which describes the item turnover frequencies, i.e. even random storage policies. Caron et al. embed the function into the formulas to calculate the expected travel distance for a layout of type D. In line with previous work they also distinguish between within-aisle and across-aisle travel. For a return routing policy and across-aisle storage the function is used to determine the farthest pick location within the aisle. For traversal routing and within-aisle storage the function influences the calculation of the probability that a pick is in an aisle close to the depot. It is also used to estimate across-aisle travel as aisles farthest from the depot have a smaller probability of being visited.

The model of Caron et al. (1998) was transferred to layouts of type A by Hwang et al. (2004). The authors first derive formulas for the estimation of travel time for a return routing policy and across-aisle COI-based storage policy. Travel time estimations are also given for traversal routing policy applying within-aisle COI-based storage. For the first time the midpoint routing policy, rather rarely considered in previous papers, is analyzed under non-random storage policies. Assuming COI-based perimeter storage the authors build upon the results for the return routing policy and estimate total travel time. For a return strategy Gudehus (2005, p. 756) extends the formulas of the original work by Kunder and Gudehus (1975) to across-aisle full-turnover-based storage policies.

Travel time models for traversal and return policies for layouts of type D are presented in Le-Duc (2005, p. 31) as well as Le-Duc and de Koster (2005b, 2005c). In contrast to previous papers (e.g. Caron et al. (1998) or Jarvis and McDowell (1991)) this new approach considers storage policies in which different product classes can be stored within one single aisle. Additionally the approach is easily transferable from layouts of type D to layouts of type F. First, the expected travel distance within a single aisle is derived by estimating the farthest pick inside the aisle. This depends on the picking probabilities of the different classes and the partial aisle length reserved for the classes. Note that the number of classes, their related probabilities and their position within the aisle is arbitrary, i.e.

any (discrete) distribution of picking frequencies on rack columns can be approximated. Due to the limited number of rack columns per aisle, the discrete approach of Le-Duc seems well suitable. The single-aisle result is then used and expanded to layouts with multiple aisles. For the traversal routing policy the correction factor for excess travel in the last aisle (see de Koster et al. (1998) and Roodbergen (2001)) is adapted to the assumption of having aisles with multiple classes.

A recent work on travel time models has been published by Sadowsky (2007). He examines return policy with single or multiple aisle visits, midpoint policy and traversal policy with or without aisle skipping in layouts of type A and C. He assumes that picks are equally distributed across the aisles. For the return policies the expected within-aisle travel is dependant on two factors: the probability of having a certain number of picks in the aisle and the location of the pick farthest from the cross-aisle. Urn models are applied to derive the first factor, resulting in a hypergeometrical distribution (note that this distribution requires the number of picks to be smaller than the number of items stored per aisle). For the second factor Sadowsky gives explicit results for exponential and uniform distributions but the formulas are able to model any continuous distribution function. The results for return policies are extended to midpoint policies. For the traversal policies, Sadowsky also uses the hypergeometrical distribution. He further compares his models with simulation and previous works, showing mostly accurate results. Unfortunately the publications of Le-Duc and de-Koster (2005) (2005b) (2005c) were not considered even though they had very similar assumptions. It would be interesting to know if there is an advantage in considering within-aisle pick frequencies by using continuous distribution functions.

Integrated Approaches

An approach focusing on traversal and return policies, random storage and layout type A was published by Rana (1990). He describes an algorithm to divide a given picking order into several trips. For each trip either traversal or return policy is used. Formulas which relate different warehouse parameters to expected travel times are given.

Galka et al. (2008, p. 253) apply the model of Sadowsky (2007) on several successive order picking zones to estimate total order picking throughput times.

Wisser (2009, p. 80) presents integrated models to calculate total costs of order picking systems. The approach analyzes and combines several existing publications on travel time (namely Kunder and Gudehus (1975), Hall (1993), Hwang et al. (2004), Jarvis and McDowell (1991), Roodbergen and Vis (2006) as well as Schulte (1996)) to derive layout parameters. Both results are then used to estimate total costs of the order picking system.

3.1.3 Approaches Using Elementary Queueing Systems

We also briefly discuss some models that apply queueing theory to order picking systems. In contrast to the previous approaches these papers consider stochastic elements like the second moment of travel time. These models are still classified as single-picker-models because they do not consider any interdependencies between multiple pickers. The influence of travel time variations thus concerns only the waiting time of an order before the picking process is started.

Bhaskaran and Malmborg (1989) present an approach to analyze the service process in a warehouse with order batching under different vehicle dispatching rules. The service time includes all necessary steps to complete a batch of orders and is estimated by a lower and upper bound, both depending on the space of the order picking area. Pandit and Palekar (1991) model warehouses of layouts B, D and E with automated guided vehicles and single command order picking. The orders arrive according to a Poisson process and the system is modeled as a system with parallel servers. The authors estimate the first and second moments of the travel time depending on the layouts. A sufficient service level which might depend on the number of vehicles can be calculated. The influence of batch size and class-based storage assignment policy was studied by Chew and Tang (1997, 1999). They calculate the average waiting time for an order or batches of orders. An estimation is given for the first and second moments of travel time for a single-block layout of type A with multiple picks per tour. The batch size to minimize turnover time is calculated for different skewnesses of item frequency. For a 2-block warehouse of layout D, random storage and order batching, Le-Duc and de Koster (2007) propose a queueing model to derive optimal batch sizes. They build on the works of Chew and Tang to determine first and second moments of travel time. The formulas are extended by two adjustment terms, considering some additional travel that stems from the different layouts. The authors conclude that the average waiting time of an order is a convex function of the batch size, i.e. a unique optimum batch size always exists.

3.2 Models with Congestion Consideration

In this chapter we will present models that have explicitly considered congestion in the analysis. In particular this means throughput does not increase linearly with increasing number of pickers. The models will be described in detail in order to give the reader some valuable insights on congestion phenomena in order picking systems. To the best of our knowledge, this section provides a comprehensive review of all models discussed in the literature.

3.2.1 Characteristics of Congestion

For a clear understanding, we will first define the different blocking situations that directly lead to congestion in an order picking system. This chapter is partly based on the descriptions given in Lüning (2005), Gue et al. (2006), Parikh (2006), Parikh and Meller (2009) as well as Furmans et al. (2009).

Waiting times in manual order picking systems are not necessarily the result of congestion and for that reason we should distinguish different situations. On the one hand, waiting times might stem from order pickers not having new orders to pick because orders are coming in slowly or staff planning was not adequately carried out. Waiting might also occur due to technical failures such as break-down of the warehouse management system or any situation resulting from act of nature. On the other hand, waiting times stem from congestion if order pickers interfere with each other when the picking process could run normally. We define the following:

In manual order picking systems we speak of congestion whenever the activities of an order picker are interrupted by another order picker while all requirements for a regular picking process are met.

A picking process is regular if orders are ready for picking, necessary supporting systems, like IT or equipment, are working and interfaces with preceding and succeeding areas are in operation (e.g. incoming goods from the receiving area, outgoing orders to a sorting system etc.).

In manual order picking systems congestion arises from different types of blocking situations. We observe those whenever two or more pickers simultaneously want to use one of the following resources:

- A rack column within the aisle and/or the space in front of the rack column
- A space of any cross aisle
- The depot and the related space in front of the depot.

We can distinguish Pick-Face Blocking, In-the-Aisle Blocking, Aisle Blocking, Cross Aisle Blocking and Depot Blocking.

Pick-Face Blocking

A situation is called *Pick-Face Blocking*, if two or more pickers want to access the same pick face, i.e. rack column at the same time. This type of blocking situation might occur in systems with arbitrary aisle width. Figure 3.2 shows how picker 2 has to wait until picker 1 has finished all picking and administrative activities at rack column i.e. pick face i.

In-the-Aisle Blocking

This type of blocking situation basically occurs in narrow-aisle systems where the aisle does not offer sufficient width for unrestricted movement

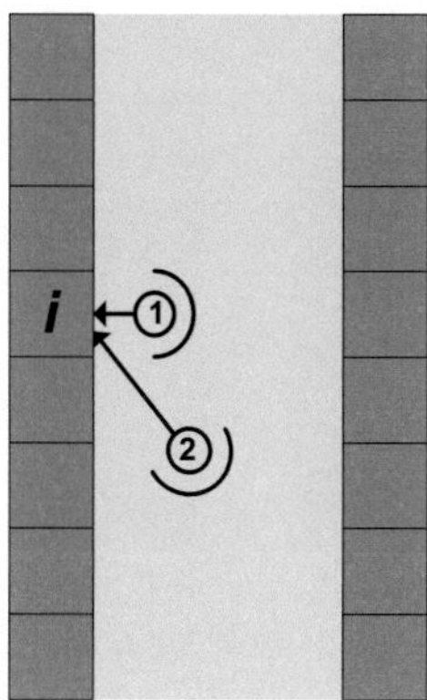

Figure 3.2: Pick-Face Blocking situation

of two or more pickers. In particular, they can not pass each other, because whenever a picker uses a certain amount of space no one else can use that space simultaneously. The direction of the picker movement defines the complexity of such situations. Figure 3.3 shows three examples. We assume the direction of movement to be fix in examples a) and b) and to be arbitrary in example c).

We identify *level-1 blocking* in example a) as picker 2 can not access his next rack column because the space in front of rack column 1 is occupied by picker 1 and aisle width restricts passing. Picker 2 has to wait until picker 1 has finished his pick and entirely left the space in front of rack column 1.

We find a similar situation in example b). In this case picker 3 is blocked by picker 2. In contrast to example a) someone who is already blocked (2 blocked by 1) causes a blocking situation by himself (3 blocked by 2). We call this situation *level-2 blocking*. Let K be the total number of pickers in the system. A *level-$(K-1)$ blocking situation* can develop. Because the direction of movement is strictly one-way in this configuration, the resulting queue can be dissolved by applying a simple first-in-first-out processing rule.

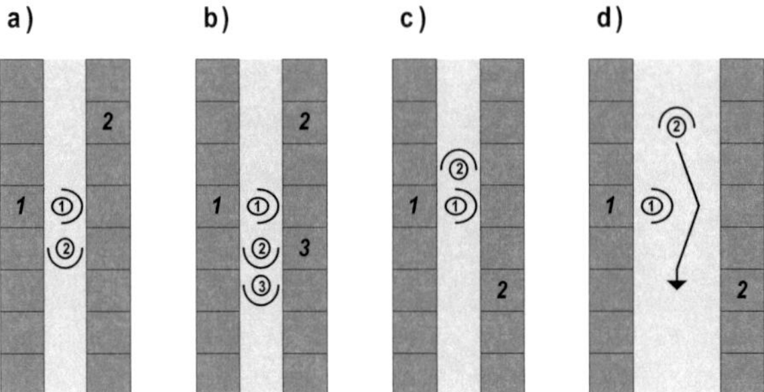

Figure 3.3: In-The-Aisle Blocking and Interference situations

Example c) shows a situation with two-way traffic. We assume that picker 1 entered the aisle from the lower cross aisle and is picking at his pick face, occupying the corresponding space. Picker 2 came from the upper cross aisle and is blocked by picker 1. We can observe a *level-1 opposite blocking situation* as soon as picker 1 has finished picking and wants to travel upwards within the aisle. In contrast to examples a) and b), such situations require priority rules defining which picker actually has to turn and walk against his original direction to clear the way.

Note that in-the-aisle blocking also covers pick-face blocking as the space corresponding to the pick face cannot be accessed in the first place thus preventing two pickers of picking from the same pick face.

In-the-Aisle Interferences

Even when aisles are wide enough to allow for passing, interferences are still possible as we might experience slower walking speeds or extra side-wards movements. Situations where pickers can pass each other at the cost of extended travel time will be called *interferences*. An example is given in d) of figure 3.3.

Total Aisle Blocking

This type of blocking situation can occur if an aisle is exclusively reserved for one picker. If a certain picker has reserved and entered the aisle all subsequent pickers have to wait until this picker has completed all picking and walking in that aisle.

Cross Aisle Blocking

Just like picking aisles the cross aisles only have limited space available and a certain part of the cross aisle can only hold one picker at a time. We speak of cross aisle blocking whenever a picker needs to use a part of the cross aisle which is already occupied by another picker. Cross aisle blocking might happen if an in-the-aisle blocking situation propagates beyond the aisle entrance.

Depot Blocking

A situation is called depot blocking if two or more pickers want to use the depot at the same time. In the worst case $(K - 1)$ of K total pickers have to wait. The complexity of such situations is manageable though as we can apply a first-in-first-out processing rule.

3.2.2 Simulation Studies

Many papers have presented simulation studies to find the best order picking system design and operating strategy for a given set of input parameters. However only few of those papers have incorporated congestion. An early study was conducted by Mellema and Smith (1988). They consider order picking truck operations in layout type A. They assume total aisle blocking and calculate throughput per man-hour and picker utilization. Ottjes and Hoogenes (1988) presented a study explicitly focusing on vehicle traffic in distribution centers. They conclude that total waiting times of order picking trucks due to congestion is increasing exponentially in the number of trucks. Pandit and Palekar (1991) studied single command operations in layout type D. They account for pick-face and in-the-aisle blocking.

A paper by Ruben and Jacobs (1999) evaluates the influence of both batch construction heuristics and storage assignment policies in an order picking system of layout type A with traversal routing policy. They specifically designed their simulation model to account for congestion. The authors give valuable insights on behavior of congested systems subject to different batching and storage policies. They showed that productivity decline is much bigger under a turnover-based storage policy compared to losses using family-based or random storage policies. One has to specifically focus on this observation if demand levels are highly variable. In cases of high demand a rather easy strategy is to increase the workforce in order to deal with the increased workload. Ruben and Jacobs have shown that family-based or random storage policies come along with more workforce flexibility compared to turnover-based storage.

A simulation study dealing with pick-face blocking was presented by Lüning (2005, p. 123). For a single aisle, assuming either random or class-based storage, central or decentral depositing as well as varying order sizes he calculates the productivity of the pickers (measured in order lines per working hour) and the corresponding relative throughput decrease. He subsumes that the productivity decrease for a rising number of order pickers is larger for class-based storage systems. He even observes that in class-based storage overall throughput is decreasing if the number of order pickers is beyond a critical value. Finally, in the scenarios studied, batching of orders is always found to be worthwhile in random storage. In class-based storage this effect is diminishing for a larger number of order pickers and negative beyond a critical size of the workforce.

Thayalan (2008, p. 41) presented a study comparing random and class-based storage policies with three routing policies (traversal, return and optimal) considering total aisle blocking. The author concludes that under class-based storage a mixture of traversal and return policy seems to do better than an optimal routing policy.

The papers of Gue et al. (2006), Parikh (2006), Parikh and Meller (2009, 2010) partly use simulation models to study the effects of congestion. Their work is mostly of analytical nature and simulation is only used for some cases. Therefore we refer to the discussion of these papers in chapter 3.2.4.

3.2.3 Combined Probability and Combinatorial Calculation

In addition to the simulation study mentioned above, Lüning (2005, p. 142) also proposed analytical methods to estimate decrease in throughput caused by pick-face blocking in one aisle. Furthermore he analyzed in-the-aisle interferences.

The pick-face blocking approach builds on probability theory and combinatorial calculations. Lüning uses the binomial coefficient to calculate the number of possible combinations of having j of K total pickers situated in front of one rack column at the same time. Weighting that with the probability that a picker is situated in front of one rack column i at a certain point in time and considering all rack columns, Lüning derives the following formula to estimate throughput decrease:

$$\lambda_{Decline} = \lambda_{OP} \cdot \left(K - \sum_{i=1}^{c} \sum_{j=2}^{K} \binom{K}{j} (p_i \cdot r)^j \right)$$

where $\lambda_{Decline}$ is the reduced throughput (order lines per hour), λ_{OP} is the throughput (order lines per hour) of a single picker without any congestion, c is the number of rack columns in the aisle, p_i is the pick frequency of rack column i and r is the probability that a picker is picking, i.e. the portion of picking time with regard to the total order picking time.

Lüning used a similar approach to estimate additional travel time caused by in-the-aisle interferences. First he derives the additional time $t_{add,K}$ a picker needs to walk in order to pass $(K - 1)$ other pickers, assuming that K pickers meet in a certain segment i. Note that one segment does not necessarily represent one rack column. After deriving the probability that K pickers interfere in segment i the number of interferences I_K is calculated. In accordance with the calculations on pick-face blocking, the binomial coefficient is used and the percentage increase in travel time X_K

is given by:

$$X_K = \sum_{j=2}^{K} \binom{K}{j} \cdot t_{add,K} \cdot I_K$$

Experiments show that for random storage travel time typically increases as much as 2% for less crowded aisles (5:1 ratio of segments to order pickers) and the increases can be as high as 7% for very crowded aisles (3:2 ratio). For class-based storage increases are much higher, namely 5% for less crowded aisles and up to 30% for very crowded aisles. Based on these results, Lüning derived some operational guidelines. In random storage there should be at most one picker per 3 meters of pick face. For class-based storage, there should not be more than two pickers per aisle, independent of the total length of the pick face as picks often tend to concentrate on a few rack columns.

3.2.4 Markov Chains and Random Walks

To quantify throughput decreases caused by congestion several authors use the concept of Markov Chains. A Markov Chain is a particular stochastic process. The system is characterized by a discrete state space and transition probabilities to get from one state to another. The underlying Markov property states that the probability distribution of future states only depends upon the current state (Waldmann and Stocker 2004, p. 5).

All subsequent models presented in this chapter aim to derive the fraction of time a picker is blocked. An order picking system with N picking locations is transformed into a closed circle as shown in figure 3.4. A picking location might be a rack column or a certain amount of space of an aisle from which two opposite rack columns can be accessed. Note that each picking location has exactly one successor, thus representing traversal routing without aisle skipping. K order pickers move along the circle in discrete time steps. At each location and time step they make a pick with probability p and no pick with probability $q = 1 - p$. p is also called the pick density. The picking time is constant and given by t_{pick}. If no pick is done the pickers walk and consume time t_{walk}. The states

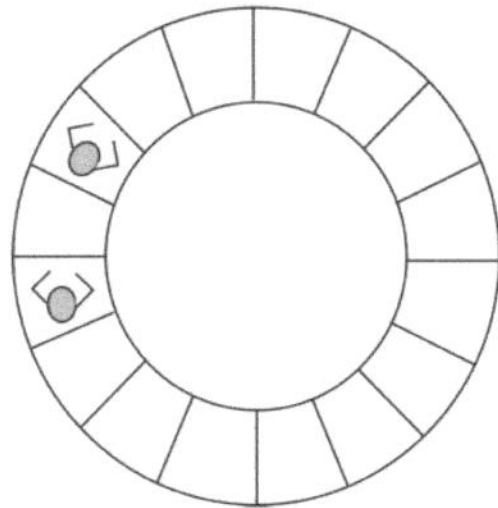

Figure 3.4: Circle representation of order picking systems (Gue et al. 2006)

of the Markov Chain are characterized by the distance between workers and the information on whether the workers picked or walked in the last time step.

Skufca (2005) considers systems with in-the-aisle blocking. He assumed infinite walking speed, hence $t_{walk} = 0$. Within a time step, a picker might thus walk until he has to make a pick or until he is blocked by another picker. Skufca derives the time fraction in which an order picker is blocked in systems with N locations and K pickers.

Building on the work of Skufca (2005), Gue, Meller and Skufca (2006) analyze an order picking system with two workers and a pick to walk ratio of $1 : 1$, i.e. $t_{pick} = t_{walk}$. They give the state space and transition matrix and derive the average percentage of time a picker is blocked in a two-picker-system as a closed-form expression depending only on p and N:

$$b_{1:1}(2) = \frac{pq}{(N-1)(p+1)^2 - 2p^2}.$$

Figure 3.5 shows the blocking time fraction over the pick density for different system sizes. The authors state that congestion is worst when approximately $0.33 \leq p \leq 0.37$. For two workers walking with infinite speed (pick to walk ratio of $\infty : 1$), the following formula is derived:

$$b_{\infty:1}(2) = \frac{1-p}{2(1-p) + (N-1)p}.$$

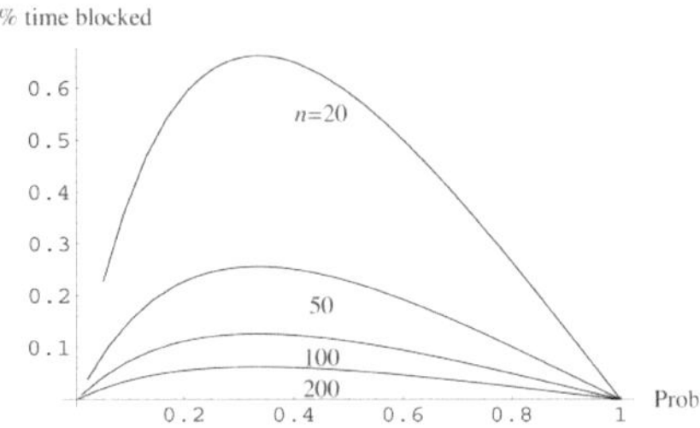

Figure 3.5: Percentage of blocked time for 1:1 pick to walk ratio and in-the-aisle blocking (Gue et al. 2006)

Blocking in the infinite speed case is proved to be higher than blocking in the 1:1 case for any scenario with $n > 2$ and $0 > p > 1$. Having obtained $b(K)$, system throughput can be estimated according to:

$$\lambda(K) = K \cdot \left[\frac{p}{pt_{pick} + t_{walk}} \right] \cdot (1 - b(K))$$

Because the model of Gue et al. (2006) cannot be easily extended to arbitrary ratios of pick time to walk time, the authors present some more findings based on a simulation model. Figure 3.6 shows the percentage of time blocked for different ratios. It appears that the critical pick density p, for which blocking is maximum, shifts to the left for increasing t_{pick}. An evaluation of random pick times revealed that congestion was slightly higher for non-deterministic t_{pick}. The major contribution of this work is the observation that congestion is highest for a critical pick density and then decreases as systems become busier.

Parikh (2006) as well as Parikh and Meller (2009) extended the use of Markov Chains. They estimate pick-face blocking in wide-aisle systems, thus allowing pickers to pass each other. They present models for the case that workers pick one item at the pick face (as was assumed in previous Markov Chain models) and the case that pickers take several items from the pick-face, leading to a higher variance in the time spent at one location.

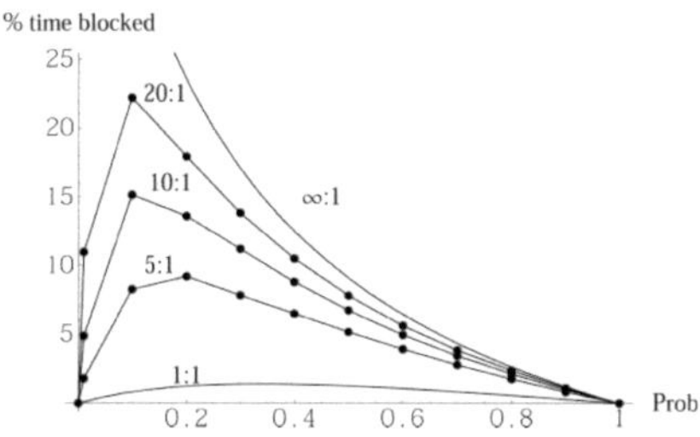

Figure 3.6: Percentage of blocked time for different pick to walk ratios and in-the-aisle blocking (Gue et al. 2006)

In line with Gue et al. (2006) the authors derive the state space and transition matrix for pick to walk ratios of 1:1. The result, again, is a closed-form expression for percentage of time blocked in a two-picker-system $b_{1:1}(2)$, depending only on the number of locations N and the pick density p:

$$b_{1:1}(2) = \frac{p^2(1-p)}{N(2-p)(1+p)^2 - p^3 - p}.$$

Analyzing the derivative, the authors state that congestion is maximum for $0.628 \leq p \leq 0.653$. Figure 3.7 shows curves of $b_{1:1}(2)$ for different systems sizes. We can observe that the curves for pick-face blocking are left-skewed, in contrast to the right-skewed curves for in-the-aisle blocking (see figures 3.5 and 3.6). The critical value of p is higher for pick-face blocking as the event of both pickers trying to access one single location at the same time is relatively rare for smaller p. As p gets close to 1, the blocking time approaches 0 as the pickers stop at almost every location. We should also note that the formulas support the intuitive assumption that systems with pick-face blocking are less congested than systems with in-the-aisle blocking (Parikh and Meller 2009, p. 238).

The analysis for pick to walk time ratios of $\infty : 1$ does not lead to a closed-form expression. For pick to walk time ratios in between the two extremes of $1 : 1$ and $\infty : 1$, blocking times were obtained by simulation. Figure 3.8 features $b(2)$ for different ratios and shows that for increasing

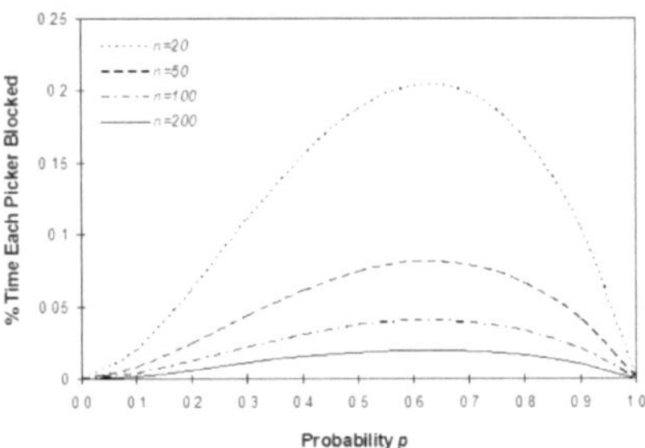

Figure 3.7: Percentage of blocked time for 1:1 pick to walk ratio and pick-face blocking (Parikh and Meller 2009)

pick times t_{pick} the maximum loss is reached for decreasing pick density p and the curves tend to be right-skewed.

Simulation is also used to determine blocking times for $K > 2$. The authors conclude that pick-face blocking should not be underestimated, especially if more than every tenth location is occupied.

Parikh (2006) and Parikh and Meller (2009) also developed models for systems were pickers pick more than one item at a pick face. The probability of picking i items at a location is described by a probability mass function depending on u and $v = 1 - u$ where v is the probability of not picking. Following the same procedure as presented above, the authors derive a closed-form expression to estimate blocking for $K = 2$ and a pick to walk ratio of $1 : 1$:

$$b^i_{1:1}(2) = \frac{u^2}{N(2 - u) - u + 2u^2}.$$

From the analytical analysis and simulation it appears that for arbitrary ratios of pick time to walk time both $b_{1,\dots,\infty:1}(2)$ and $b_{1,\dots,\infty:1}(K)$ increase monotonically for increasing pick densities.

We should add that the variance of t_{pick}, stemming from the random number of picks at one pick face, can have a significant influence. In one particular scenario where i items had to be picked from $N > i$ locations,

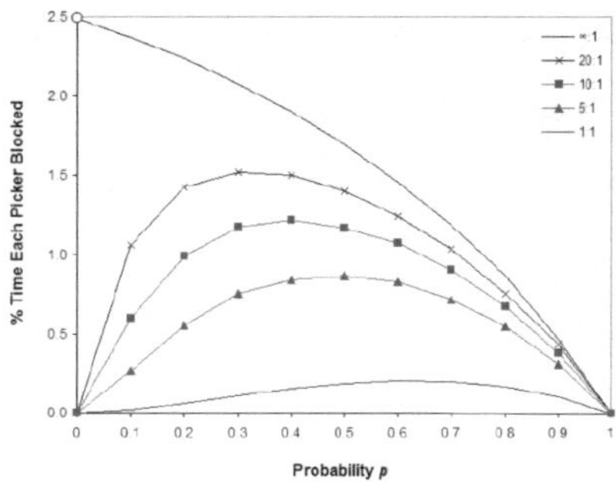

Figure 3.8: Percentage of blocked time for different pick to walk ratios and pick-face blocking (Parikh and Meller 2009)

the portion of blocking increased from 2.5% (all items picked from different locations) to 9% (some items picked from the same locations) (Parikh and Meller 2009, p. 245). Parikh and Meller (2010, p. 400) conclude that in comparison to deterministic pick times congestion is more pronounced for non-deterministic pick times. Furthermore, congestion might have a big influence in systems with high pick densities.

3.2.5 Queueing Networks

As we explained earlier (see chapter 3.1.3), some authors have used queueing models for the analysis of manual order picking systems, but these approaches had no consideration of congestion as we defined it (see chapter 3.2.1). The research on this field is very limited to say the least.

Open queueing networks with finite buffers have been used by Pan et al. (2005) as well as Pan and Shih (2008). They considered a layout of type A with ν aisles and a traversal strategy without aisle skipping. The model assumes total aisle blocking and a buffer is located in front of each aisle. The aisles are represented by one single queueing system, thus the system of ν aisles is transformed into an open queueing network of ν tandem queueing systems with finite buffers each. Service times,

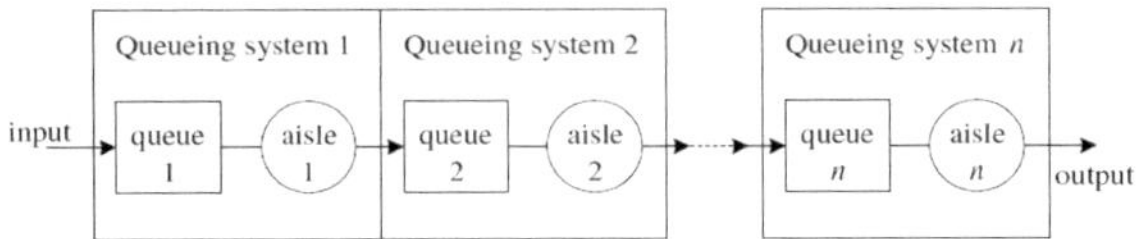

Figure 3.9: Open queueing network of an order picking system with ν aisles (Pan and Shih 2008)

which include both travel and picking, are assumed to be exponentially distributed. Let $P_{i,j}$ be the probability that an item stored in rack j of aisle i (which has h racks) is picked in an order, w be the total width of an aisle (including depths of the opposing racks) and l be the total aisle length. Then the expected distance traveled within aisle i is given by:

$$E(D_i) = w + l \cdot \left(1 - \prod_{j=1}^{h} (1 - P_{i,j}) \right).$$

The expected time to pick items in aisle i is defined as:

$$E(K_i) = t \cdot \sum_{j=1}^{h} P_{i,j}.$$

The service rate μ_i is then easily determined as:

$$\mu_i = \frac{1}{E(D_i) + E(K_i)}.$$

As this service rate does not incorporate any blocking situations caused by finite buffers, the authors use the approximation method of Takahashi et al. (1980) to estimate effective service rates, which are given by:

$$\frac{1}{\mu_i^*} = \frac{1}{\mu_i} + \frac{\frac{\rho_i^s(1-\rho_i)}{(1-\rho_{i-1}^{s+1})}}{\mu_{i+1}^*}.$$

Throughput values are then calculated by treating each queueing system as an $M|M|1$ system and applying μ_i^*. Pan and Shih (2008, p. 386) conduct an experiment comparing random storage policy to class-based storage policy and summarize that random storage yields higher throughput as the picking area is more uniformly utilized.

Closed queueing networks were first used by Furmans et al. (2009) and are the basis for the modeling approach presented in this thesis. We refer to chapter 4 for a detailed description.

3.2.6 Related Non-Order Picking Approaches

Material handling systems are closely related to order picking systems and might also suffer from congestion. Gudehus (1976) has discussed estimations of queue lengths and waiting times in front of conveying elements. He also addressed the tailback probability in a sequence of work stations in (2005, p. 517). Schmidt and Jackman (2000) have modeled cyclic continuous conveyors as an open queueing network. Bartholdi and Gue (2000) used queueing theory to design efficient layouts for crossdocking terminals. Tempelmeier and Kuhn (1993, p. 141) consider flexible manufacturing systems and estimate the impact of material handling congestion by means of closed queueing theory. A general method to measure congestion in material flow systems was presented by Faißt and Lippolt (2002).

Quadratic Assignment Problems are often used in layout and facility design. Chiang et al. (2002, 2006) have incorporated congestion and workflow interference in this subject. Zhang et al. (2009) combine probabilistic models and flow problems to model and minimize workflow congestion and give an extended literature review.

Apart from intralogistic flows, we can observe congestion almost daily in everyday life. Weber and Weiss (1994) have examined the cafeteria problem, where people move along a line of M stations and receive service. The movement is modeled as a Markov Chain and resulting queues and throughput are derived. A big amount of research exists on traffic systems. As macroscopic traffic models do not consider the individual elements of traffic systems, microscopic traffic models seem to be more interesting with regards to order picking systems. A limited number of approaches seem partly transferable, e.g. the cellular automaton proposed by Nagel and Schreckenberg (1992). An extensive literature review on vehicular traffic flow modeling is given by Hoogendoorn and Bovy (2001).

3.3 Chapter Conclusion

The models presented in chapter 3.1 do not incorporate congestion. However, numerous of these papers have given an indication that congestion might influence the travel time and identified a future research activity in the development of new models.

The simulation studies have provided some very useful indications on how congestion can affect throughput. Depending on the phase of application, e.g. rough or detailed planning, the approach of simulation has some disadvantages (Bolch et al. 1998, p. 609). A simulation model is very time consuming and needs to be run with a certain number of replications in order to attain stable results. A simulation experiment requires a more or less complex programming of processes that usually leads to a large amount of source code. Transparency decreases with larger models. The level of detail provided by simulation might not be needed in a rough planning phase where the focus is on the quick analysis of numerous alternatives. It would therefore be helpful to have analytical methods for the analysis of manual order picking systems with congestion.

Lüning (2005) provided some valuable models and findings on the possible productivity decrease caused by congestion, although the applicability to more diverse order picking systems seems to be rather limited. The fact that only a single aisle is considered restricts the approach to zone picking configurations. The interdependencies stemming from having several aisles in the system and thus applying different routing strategies can not be incorporated into the analysis. The picking times are considered deterministic, thus stochastic influences are left out. The approach is limited in terms of blocking situations with the first model considering pick-face blocking and the second model considering in-the-aisle interferences. Furthermore, the modeling of in-the-aisle interferences allows for K pickers to simultaneously pass each other. Depending on the width of the aisle such situations probably are not possible in real systems.

Markov Chain models have greatly improved understanding the implications of blocking situations in order picking systems. However, models appear to be able to analyze rather specific systems only. First, the $1 : 1$ and $\infty : 1$ walk to pick time ratios used in the analysis will most likely not be found in real-life systems. Secondly, the analytical expressions mostly can consider only two order pickers and simulation is used to consider systems with more order pickers. A formula is only given for the $\infty : 1$ case. Using a closed circle to represent the arrangement of locations only allows to evaluate the traversal routing policy without aisle skipping. Other routing strategies involving any kind of branching can not be analyzed. Furthermore all locations have the same probability of having a pick, thus storage policies like turnover-based or class-based storage can not be incorporated. Finally, times to pick one item are mostly assumed to be deterministic and equal for all locations. Thus it is not possible to differentiate between different system elements, e.g. depot or rack columns with light or heavy items, which might require different times for processing.

The application of open queueing networks with finite buffers is an interesting approach. Nevertheless, some open questions on the assumptions remain. Because of total aisle blocking we cannot analyze situations with multiple pickers being located within an aisle simultaneously even though such situations surely occur in real-life systems. In open queueing networks, the number of pickers is arbitrarily high or low and variable over time. In particular this leads to the assumption that an unlimited number of pickers is available at the depot. The sequence of queueing systems, i.e. aisles, is strictly specified, enabling only the analysis of traversal policy without aisle skipping. The assumption of exponential service times is debatable, as very low times have the largest probability. In real systems however, we should consider a minimum walking time in the aisle and each pick will require a minimum time for administrative activities. Detailed discussions on the distribution of service times in manual order picking systems are presented in chapter 4.7. Finally, concerning the work of Pan and Shih (2008), we claim that the expected distance within an aisle $E(D_i)$ is fix for traversal strategy without aisle skipping and should not be dependent on $P_{i,j}$.

The approaches proposed for material handling systems are based on queueing theory and seem promising. At the same time they are subject to some restricting assumptions such as the type of network used or the exponential service times. Models of layout and facility design as well as traffic do not provide the necessary level of detail to model the process flow of an order picking system, e.g. the defined cyclic movement starting and ending at the depot, the application of routing strategies or picking times.

Having critically evaluated the existing throughput models we are able to identify situations in which these models are not applicable. Based on this, we can formulate the requirements for a new method to estimate throughput in manual order picking systems with congestion consideration:

- The method should be of analytical nature in order to allow for the quick analysis of many different alternatives.

- The method should enable the analysis of multiple-picker systems without pre-defined restrictions concerning the number of pickers.

- The method should consider different types of blocking situations, especially pick-face, in-the-aisle, cross aisle and depot blocking situations.

- The method should be adaptable to different layouts, i.e. different lengths and widths of order picking systems. In particular, the approach should be able to analyze multiple-aisle systems.

- The method should allow some flexibility concerning the route of a picker. In particular, the sequence of visited rack columns is not necessarily the same for two successive orders.

- The method should consider random and non-random storage policies. Thus the items of different rack columns might have different probabilities to be on an arbitrary order.

- The method should consider stochastic picking times. In particular, different rack columns might have different picking time means and variances. The type of distribution for these picking times should be general to enable a realistic modeling.

In the following chapters, we will present models of closed queueing networks that fulfil these requirements and hence provide a method to estimate throughput in manual order picking systems with congestion consideration.

4 Queueing Models of Manual Order Picking Systems

We will review the necessary basics of queueing theory and briefly present our motivation to use this methodology in chapter 4.1. Subsequently, we will conduct a step-by-step transformation of the parameters defining a manual order picking system into the parameters of a queueing model. We start with the basic assumptions of the order picking system and the choice of queueing network type in chapter 4.2. Necessary elements (e.g. aisles, racks, depot) are considered in chapter 4.3. The one-way traversal routing policy will be introduced in chapter 4.4. Transition probabilities implementing this routing policy are given in chapter 4.5 for random storage and chapter 4.6 for class-based-storage. Finally, we will consider the order picking times (e.g. walking, searching, picking) in chapter 4.7.

4.1 Essential Basics of Queueing Theory

4.1.1 Classification, Benefits and Limits

Queueing theory can be used for the performance evaluation of telecommunication networks, production lines and material flow systems. In principle, it expands an analysis of systems with a single element to systems with multiple interacting elements. When analyzing the average time spent at a supermarket cash desk the stochastic elements in a system with a single customers would be the number of items in the shopping cart and the speed of the cashier. Then an average time needed for scanning and payment of items can be calculated. However we might also experience waiting times as other people are already at the cash desk in the instant of our arrival. Queueing theory enables to consider both service and waiting times in a system with multiple customers.

By means of simulation a system analysis can be conducted on any level of detail, apparently including various stochastic elements (Bolch et al. 1998, p. 1). In early planning phases, which are characterized by a large number of design alternatives, the use of simulation might lead to excessive effort both in model building and experimenting. As a result simulation is often used in detailed planning phases after a set of alternatives has been pre-selected for further analysis (Rall 1998, p. 17).

Queueing theory is an eligible tool for pre-selection of alternatives as it incorporates statistical input variables. Thus queueing models consider interdependencies stemming from the stochastic behavior of different system components. At the same time, the calculations are of analytical nature so many different alternatives can be evaluated in relative short amount of time. The calculated performance measures usually give a good estimation of the underlying system's performance even if some assumptions are violated (Suri 1983). Furthermore the input variables can easily be manipulated in order to analyze system performance in different scenarios to evaluate the robustness of an alternative (Furmans 2000, p. 2).

Despite the capability to consider stochastic input parameters, continuous time queueing theory can not explain the full range of potential influences stemming from a random variable. For example, different skewnesses or multiple modes of a distribution will not affect the results[1]. The simplification of a real-life system as modeled by queueing theory mostly goes beyond that of a simulation model, asking for the planner's ability of abstraction to be higher. Historically, queueing theory has been used in telecommunications and processor networks. In the last two decades tough, many problems of material flow systems have been studied by means of queueing theory (see Furmans (1992, p. 5, 2000, 2004) as well as the references presented in chapter 3).

---

[1]Queueing theory in the discrete time domain can overcome this disadvantage. As such models are not considered in this thesis, we refer to Furmans (2004) or Schleyer (2007)

4.1.2 Single-Node Queueing Systems

An elementary queueing system consists of one or multiple identical servers and a buffer (or waiting room) located in front of the server(s). Customers (also called elements or jobs) arrive at the system and receive immediate service if at least one server of the system is available (or idle). If all servers are busy, the arriving customer is buffered. When a server becomes available one customer is selected for service according to the queueing discipline (Bolch et al. 1998, p. 209). In order to fully define an elementary queueing system, we need to know the following characteristics (Gross and Harris 1998, p. 3):

- The arrival stream to the system can be described by the distribution of interarrival times. It is commonly assumed that the sequence of interarrival times is a set of independent and identically distributed random variables. The average interarrival time between successive customers $E(t_a)$ can be used to calculate the average arrival rate λ:

$$\lambda = \frac{1}{E(t_a)}.$$

- The service process can be characterized by the sequence of service times, which again is a set of independent and identically distributed random variables. The average service time of one customer is denoted by $E(t_s)$ and is used to calculate the average service rate μ:

$$\mu = \frac{1}{E(t_s)}.$$

- The number of servers m describes the number of parallel identical servers. As a consequence, at most m customers can be serviced at a single point of time.

- The capacity of queueing system i is denoted by B_i. It is defined as the sum of the buffer spaces and the number of servers. Models commonly assume an infinite capacity of the buffer, hence $B_i = \infty$. However, many real-life systems have buffers of finite size only (Balsamo et al. 2001, p. 15). We speak of zero-buffer-systems, if $B_i = 1$.

- The queueing disciplines most frequently used are First-In-First-Out (FIFO), Last-In-First-Out(LIFO) or Processor Sharing (PS). Other disciplines are described by Bolch et al. (1998, p. 211).

Following the notation of Kendall (1953), we can describe elementary queueing systems by a set of five variables:

$$A/S/X/Y/Z$$

where A characterizes the arrival process, S the service process, X the number of servers, Y the system capacity and Z the service discipline. For A and S, the following symbols are typically used: M for exponential distributions, D for Dirac, i.e. strictly deterministic processes, E_k for Erlang distributions with k phases, C_k for Cox distributions with k phases and G for general distributions for which first and second moments are known.

Depending on the characteristics of the system, the literature offers methods to calculate several performance measures. These include the utilization ρ which denotes the fraction of time in which the server is occupied. λ refers to the system throughput, which is defined as the number of customers served in a single time unit. In the case of infinite buffers λ equals the arrival rate while for finite buffers the throughput can be smaller than the arrival rate. Other key figures include the waiting time t_w, the throughput time (or sojourn time or lead time) t_v, the mean number of customers in the queue N_W and the mean number of customers in the system N_S. For a discussion on calculating performance measures we refer to Kleinrock (1976), Bolch et al. (1998), Gross and Harris (1998) or Furmans (2000) with a focus on queueing theory in material flow systems.

4.1.3 Queueing Networks

We speak of queueing networks whenever at least two elementary queueing systems are connected and customers can be transferred between some of these nodes (Bolch et al. 1998, p. 263). A network consists of N elementary systems, each with parameters μ_i, m_i and B_i. We can distinguish two main classes of queueing networks.

In *open queueing networks*, customers can enter and leave the system from/to external sources/drains which lie beyond the system limits. Let λ_{0i} be the arrival rate of customers from an external source to system i and $q_{j,i}$ be the probability that a customer after finishing service at system j is transferred to system i. Then we can define the traffic equations as follows:

$$\lambda_i = \lambda_{0i} + \sum_{j=1}^{N} \lambda_j \cdot q_{j,i}, \text{ for i = 1,...,N.}$$

It is important to know that in open networks, the external arrival rates λ_{0i} act as an input parameter and the total number of customers in the network K is a resulting value, which can vary over time.

In *closed queueing networks*, no customer can enter or leave the system. As a consequence, the number of customers K in the network is constant and the throughput λ is a resulting value. Another important parameter in closed networks is the visit ratio e_i. It indicates the relative portion of network throughput that goes through system i. We use the transition probabilities $q_{j,i}$ to derive e_i:

$$e_i = \sum_{j=1}^{N} e_j \cdot q_{j,i}, \text{ for i = 1,...,N.}$$

As there are only $(N-1)$ independent equations, we usually define $e_1 = 1$.

Many publications exist on how to calculate performance measures for queueing networks. Important contributions include the works of Jackson (1957), Gordon and Newell (1967) and Whitt (1983). Useful overviews are given by Bolch et al. (1998) and Furmans (2000).

4.1.4 Queueing Networks with Blocking

In queueing networks, the phenomenon of blocking, i.e. congestion, can occur if a queueing system i has a finite capacity B_i. Then the queue length of system i can directly influence the processes at any preceding node j with $q_{j,i} > 0$. If system i is full, i.e. all spaces of the buffer are occupied, system j might become blocked (Balsamo et al. 2001, p. 28).

The use of finite buffers allows for the analysis of a wide range of real-life problems. However, blocking networks are also more complicated to solve as the service process of an elementary system depends on the current states of its succeeding systems (Onvural 1993).

The different mechanisms of a blocking situation are represented by three main types of blocking protocols: Blocking After Service (BAS), Blocking Before Service (BBS) and Repetitive Service Blocking (RS)[2] (Onvural 1990).

Under BAS, system j is said to be blocked if a customer bound for system i cannot continue because no buffer space is available at i. The customer will remain in the server of system j and block subsequent customers. Under BBS, a customer will only be processed in system j if the destination i is not full. If system i has available buffer spaces upon the beginning of service in system j but then becomes full, the service at j interrupts and we usually assume that the amount of service received thus far is lost. For further discussion of BBS, see Balsamo et al. (2001). Finally, under a RS protocol a customer that has finished service at j and finds destination system i full will return to the rear end of the queue at system j and thus has to wait for all customers of j to finish their service.

Research of closed queueing networks with blocking has been focusing on approximation methods as due to the complexity, the exact calculation of performance figures is often infeasible. Most algorithms are configured to specific network configurations, assumptions concerning the service time and the type of blocking protocol. A majority of papers has dealt with cyclic (i.e. each node has one fix successor) networks assuming exponentially distributed service times and BAS protocol. Our model of manual order picking systems with congestion consideration will lead to certain assumptions for which only very few applicable algorithms remain. We will present those in chapter 5.1. For descriptions of those algorithms not applicable we refer to Perros (1989, 1994), Onvural (1990), Dallery and Kouvatsos (1998) as well as Balsamo et al. (2001).

---

[2]According to Balsamo (1993), the following terms are also used: BAS: type-1, transfer, manufacturing, non-immediate and classical blocking. BBS: type-2, immediate and service blocking. RS: type-3, communication, repeat and rejection blocking.

4.1.5 Motivation for the Use of Queueing Theory

The use of queueing theory for the modeling of order picking systems with congestion consideration is motivated by the following facts:

- Many parameters of order picking systems are of stochastic nature, e.g. picking times or the sequence of visited aisles and racks.

- The walk or ride of a picker through the system can basically be considered as a sequence of visiting stations where a certain kind of service, e.g. picking or walking, is conducted.

- Pickers are usually not working in the system alone but share the infrastructure of the system, e.g. aisles, cross aisles, depot, with other pickers.

- The combination of stochastic elements and multiple-picker operations can lead to waiting times if two or more pickers want to use the same infrastructure at the same time, e.g. two pickers simultaneously wanting to pick from a certain rack aisle or two pickers simultaneously wanting to use the same segment of a narrow aisle.

- As multiple pickers cannot be served at the same time, at least one picker has to wait for another picker to finish the service, i.e. the picker has to enter some kind of waiting room.

We conclude that the movement of pickers through an order picking system is relatively similar to the movements of customers through a network of queueing systems.

4.2 Assumptions on the Order Picking System and Choice of Network Type

Assumptions on the Order Picking System

The manual order picking system consists of an even number of aisles ν. We assume that the two racks belonging to one aisle have the same length and the same number of rack columns such that there is symmetry along the center line of the aisle. A single rack column is assumed to be $1m$ long and $0.5m$ deep. The order picking system is L meters long and W meters wide. We can characterize the system's layout by the warehouse shape factor WSF, which is the ratio between L and W. The pickers are either equipped with picking carts or trucks and walk or drive through the system. At the depot they receive an order with n picks (or order lines). We assume that there always is a sufficient number of orders waiting at the depot, i.e. pickers never have to wait for orders. There is an administrative time for each picker to deposit a finished order and receive a new one. The picker stops in front of any location with picks. At these locations there will be an idle time to prepare for the picking activity. Additionally there will be the picking time which includes grabbing an item and putting it onto the picking cart or truck. In terms of storage location assignment policy, we will distinguish between random storage and within-aisle class-based storage. The sequence of visiting the picking locations is determined by the routing policy. We consider a one-way traversal policy with or without aisle skipping. There will be K pickers working in the system simultaneously. The aisles are narrow, thus pickers cannot pass each other within an aisle. Aisles are $1m$ wide and we assume that pickers potentially experience in-the-aisle[3], cross aisle as well as depot blocking. The number of picks n may or may not be the result of an order batching procedure. In the following we will not include zoning policies and consider the system as one single zone. Figure 4.1 shows an example of the underlying order picking system with $\nu = 10$ aisles and $K = 5$ pickers.

Choice of Network Type and Relevant Parameters

To develop a queueing model of the system, we need to specify the type of network first. It is quite obvious that pickers will be represented by the moving customers of the queueing model. As a matter of fact, a picker represents an order that is carried through the system. All other resources (depot, aisles, cross aisles) have a fix location and will therefore be modeled by queueing systems.

The choice of network type is based on two important observations. First, each picking tour will ultimately begin and end at the depot. We can

---

[3]Note that this includes pick-face blocking.

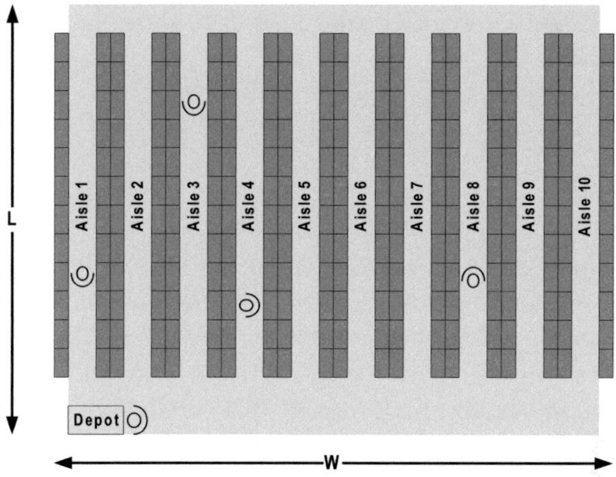

Figure 4.1: Layout of a manual order picking system

think of this as pickers moving through a circle with some possibilities for shortcuts. Secondly, the pickers do not enter or exit the system but stay within the system for a certain period of time.

With regards to the remarks in chapter 4.1.3, we model the order picking system as a *closed* queueing network. We can thus reproduce the circular movements and have a fixed number of pickers. Had we used an *open* queueing network, we would have been able to model circles but there would have been external sources and drains, leading to a variable amount of pickers in the system. This would have represented a situation where an incoming order is immediately served by a picker. There are attempts to have a flexible workforce serving different areas of the warehouse depending on the areas' workloads. However, an order-explicit availability of pickers seems to be out of reach.

In order to calculate characteristic values of a closed queueing network, several input parameters have to be derived:

- N elementary queueing systems representing the resources of the system with their respective number of servers m_i and the system capacity B_i. Resources were identified above (see chapter 3.2.1) and include:

- Rack columns and aisles
- Cross aisles
- Depot

- K customers representing the pickers.

- Visit ratios e_i which are calculated by using the transition probabilities $q_{j,i}$ and setting $e_{Depot} = 1$ as each picker will pass by the depot on each trip. By altering the visit ratios, we can model the traversal routing policy with or without aisle skipping.

- Average service time and service time variability characterizing the times needed at each queueing system.

4.3 Modeling of Resources

In the following we will derive the number of elementary queueing systems N, their respective number of servers m_i and the system capacity B_i.

4.3.1 Resources within an Aisle

An aisle is made up of two opposing racks or shelves and pickers travel along the center line of the aisle. We divide each aisle into a certain number of segments and assume that a segment covers the space in front of two opposite rack columns. An aisle offering a total of 20 rack columns consequently will be divided into 10 segments, each directly corresponding to two opposite rack columns. A picker uses a segment to either walk or drive and might also stop to do a pick at the respective rack columns. In terms of queueing theory both walking and picking can be considered as services which a picker receives at a certain segment. We will denote the number of segments per aisle with h. Each segment will be modeled by an elementary queueing system.

As we assume narrow aisles, one segment can be used by only one picker at the same time. It follows that $m_i = 1$ for all queueing systems, because no more than one picker can receive service simultaneously at one segment. It also follows that $B_i = 1$ because if a segment offers space for only one picker there is no buffer between neighboring segments. Let i be an elementary queueing system representing a certain segment. Then i itself is the buffer of the succeeding queueing system $i + 1$ and respectively queueing system $i - 1$ is the buffer of queueing system i. This logic implements the Blocking-After-Service protocol as a picker in system i is blocking the server of system i as long as system $i + 1$ is occupied. The following statement summarizes the transformation:

In an aisle with 2h rack columns we can identify h segments, each corresponding to the space in front of two opposite rack columns. Each individual segment of an aisle is represented by a queueing system with one server ($m_i = 1$) and a buffer size of 0 ($B_i = 1$). An aisle is then made up by a straight sequence of h zero-buffer queueing systems.

Figure 4.2 visualizes the transformation of a single aisle into a queueing model.

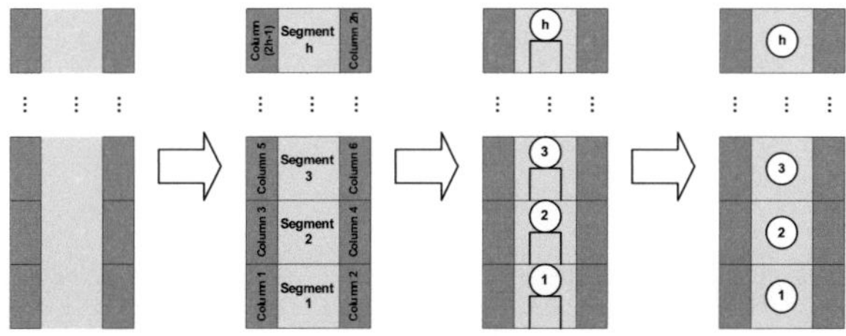

Figure 4.2: Queueing Model of a single aisle

4.3.2 Resources of a Cross Aisle

A picker uses a cross aisle to switch between adjacent aisles or to skip aisles. We assume that no racks are located at cross aisles, thus the pickers solely uses them for walking. We first consider the *lower cross aisle* and the adjacent *return path*, which are both located on the depot-side of the order picking system. The return path begins right after aisle ν and is used as soon as a picker exits an even aisle with no picks left. Like before, we divide the cross aisle into segments, which can be used by only one picker at a time. Each segment will be modeled by an elementary queueing system with one server and no buffer.

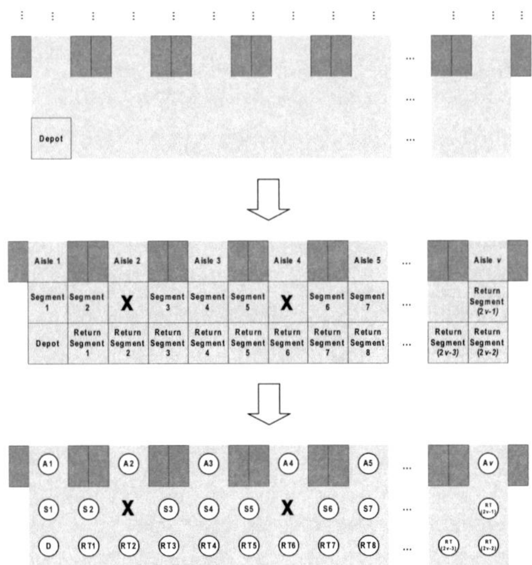

Figure 4.3: Queueing model of the lower cross aisle

The segments marked with an x denote junctions where two travel paths cross. We do not model these junctions with queueing systems for two reasons. First, we want to allow a picker coming from an even aisle to access the return path even when there is a queue on the lower cross aisle. Secondly, this approach hugely facilitates the calculation of transition probabilities (see chapters 4.5 and 4.6) while coming along with only minor modeling inaccuracies. Figure 4.3 shows the structure of the queueing model representing the lower cross aisle and return path.

The transformation of the *upper cross aisle* is straightforward as picker paths do only branch and merge but never cross. Thus the total space of the upper cross aisle can be divided into segments which are then modeled by elementary queueing systems with one server and no buffer. Figure 4.4 visualizes the transformation.

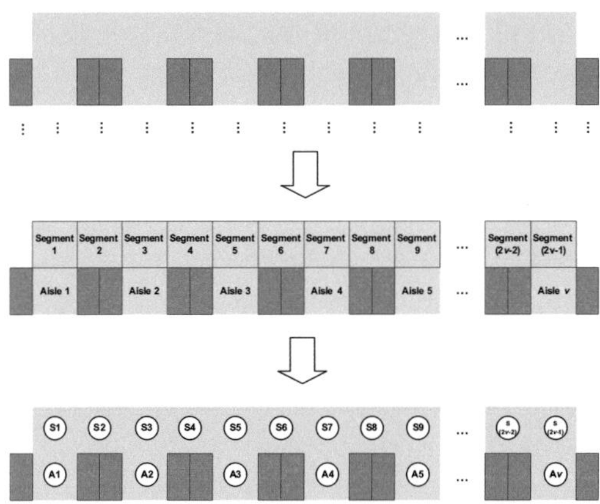

Figure 4.4: Queueing Model of the upper cross aisle

Note that these rules result in feasible transformations no matter whether pickers walk or drive. We assume that the space covered by a picker and his device is typically larger than the depth of a rack column. This assumption seems reasonable when pickers push their picking carts as well as when they drive a fork lift truck. Hence, we model the segment corresponding to the front face of two racks by one elementary queueing system with $m_i = 1$ and $B_i = 1$.

4.3.3 Depot

The depot is a work place which occupies a certain amount of space. It will be considered as one segment, which can hold one picker at a time. Therefore we will model it as an elementary queueing system with one server and capacity $B_{Depot} = 1$. As there is no buffer, the queue can reach back onto the return path.

We assume the depot to be located at the lower left corner of the system (see figure 4.1). Several studies (e.g. Roodbergen and Vis (2006)) show that the center position is best in terms of travel time. However, the authors also conclude that the location does not have a significant influence. Moreover, we place the depot on the left side as this simplifies the realization of one-way traffic, which is an important feature in narrow-aisle systems. We describe in chapter 4.4 how such rules ensure picker flow without the need for priority rules to resolve any blocking situation.

4.3.4 Overall Number of Resources

Based on the presented transformation rules for different types of resources, we can calculate the total number of queueing systems in the network:

$$N = \nu \cdot (h + \frac{11}{2}) - 3 \qquad (4.1)$$

4.4 Introduction of One-Way Traffic Rules for Traversal Routing

In narrow-aisle systems, the fact that pickers cannot pass each other is basically not complicated if order pickers travel in the same direction (see figure 3.3 a/b). However, if we apply the original traversal policy, pickers can travel the same aisle in different directions depending on the number of aisles skipped beforehand. To clear opposite blocking situations we would need priority rules defining which picker has to make way for the other. One picker has to turn around and walk against his original direction. Once the prioritized picker exits the aisle, the non-prioritized picker can retry traversing the aisle.

We identify the following problems for such priority rules. First they would likely be based on typical criteria like earliest due date or smallest slack time. This data is order-specific and thus situation-specific. In many systems, due to the lack of equipment or necessary IT, this information will not be available at all in the moment of blocking. Furthermore, in rough planning phases this level of detail is not available. Secondly, depending on the number of pickers the turn-around process itself may cause new and even more complicated blocking situations. Pickers might often move in zig-zag-patterns between aisle entrance and blocking location, causing additional travel. We will not be able to identify any kind of flow and the complexity to control and operate such a system will quickly increase.

To avoid these situations we define one-way rules, such that aisles and cross aisles may only be traveled in one direction. Therefore we need to re-define the original traversal strategy as described in chapter 2.3.2. We will refer to the new strategy as the *one-way traversal strategy with aisle skipping*:

- Travel from the lower cross aisle to the upper cross aisle is allowed in odd aisles only.

- Travel from the upper cross aisle to the lower cross aisle is allowed in even aisles only.

- The first two rules imply that a picker, when located on the lower cross aisle, has to travel an odd aisle before picking in an even aisle. Consequently, when located on the upper cross aisle, the picker has to travel an even aisle before picking in an odd aisle.

- When deciding to enter an aisle the picker only considers potential picks in the next two aisles, i.e. there is no predictive aisle traveling.

- Pickers are allowed to skip aisles but only if there are no picks in at least the next two aisles.

- As the depot is arranged in front of aisle 1, travel between adjacent aisles has to be from left to right.

- Only if pickers have finished all their picks they are allowed to use the return path to travel from right to left. The return path is located below the lower cross aisle.

With these rules we eliminate any kind of opposite blocking. Figure 4.5 shows an example for one-way traversal routing.

It also illustrates a disadvantage of this strategy: the order picker has to traverse aisle 3 even tough there is no pick in that aisle. He has to enter aisle 3 because the pick in aisle 2 forced him to move to the lower

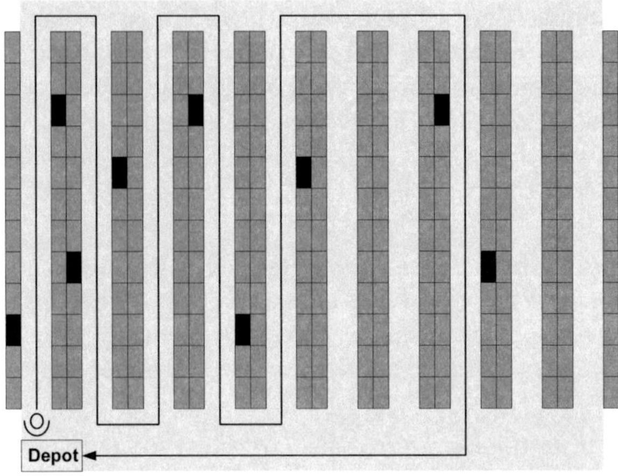

Figure 4.5: One-way traversal routing policy

cross aisle. By definition, even aisles have to be entered from the upper cross aisle tough and aisle 3 offers the only possibility to switch cross aisles before making the pick in aisle 4. Had there been no pick in aisle 2, the picker could have taken the upper cross aisle to skip aisles 2 and 3. The additional travel associated with this rule has to be accepted in order to avert complicated oppositional blocking situations. The extent of additional travel for one-way traffic will strongly depend on the ratio between the number of picks n and the number of aisles ν. For big ratios, i.e. many picks per aisle, the additional travel tends to disappear, as it will become likely that all aisles have to be entered anyway.

4.5 Modeling of One-Way Traversal Routing for Random Storage Policy

The routing method in our queueing model is represented by the visit ratios e_i and the transition probabilities $q_{j,i}$ respectively. The latter incorporate both the structure of the order picking system and the specific rules of the routing policy as they explicitly define which system i possibly follows system j. Some movements are predefined and some depend on the number of aisles ν, the number of segments per aisle h, the number of order lines n and the storage policy. In the following, we will refer to the aisle number as α with $\alpha = 1, 2, 3, ..., \nu$.

4.5.1 Predefined Movements

The aisle-based structure of the underlying order picking system and the rules specified by the one-way traversal routing policy facilitates the derivation of $q_{j,i}$ as a large amount of transitions is well defined a priori. Within an aisle, all queueing systems are arranged in strict sequence and one-way traversal routing implies that $q_{j,i} = 1$. At the last segment of each odd aisle α_{odd}, we require the picker to turn right onto the cross aisle. When skipping aisles by using the upper cross aisle a picker is not allowed to enter an odd aisle α_{odd} due to one-way restrictions. Likewise, a picker skipping aisles by using the lower cross aisle can not enter an even aisle α_{even}. Finally, we know that once the picker enters the return path, he has to travel to the depot straightaway.

4.5.2 Decision Points for Non-Predefined Movements

A few locations remain where the picker has to choose his path and thus possibly $q_{j,i} \neq 1$. We will call such a location a *decision point*. The calculation of those $q_{j,i}$ is rather not straightforward. This is because the decision on whether to enter an aisle does not only depend on the pick frequencies within the next two aisles but also on the picks done in proceeding aisles and also on the distance to the very first aisle $\alpha = 1$.

For an order picking system with one-way traversal strategy as shown in figure 4.6, we can distinguish and characterize three types of decision points:

- Decision Point A_x: at these segments of the upper cross aisle a picker has to decide whether to enter the next even aisle α_{even} or continue along the cross aisle in order to skip the next two aisles.

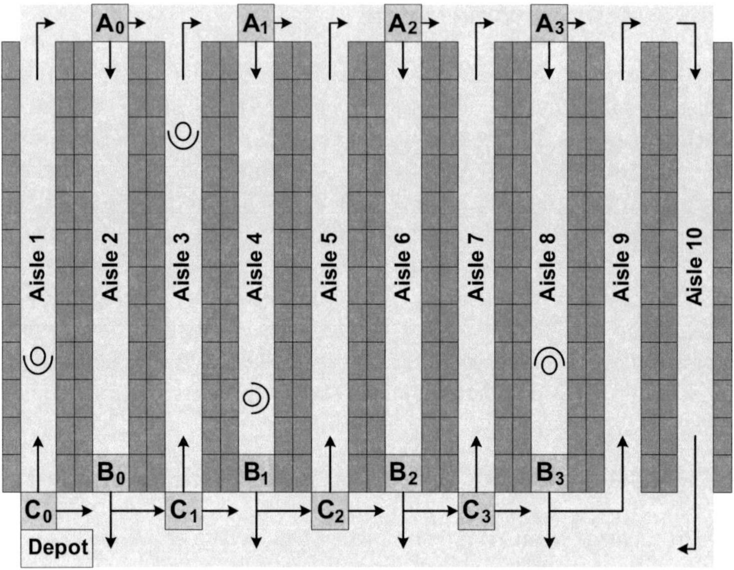

Figure 4.6: Different types of decision points

- Decision Point B_x: at the last segment of an even aisle α_{even} the picker has to decide whether to continue towards the cross aisle or continue towards the return path and the depot respectively.

- Decision Point C_x: at these segments of the lower cross aisle a picker has to decide whether to enter the next odd aisle α_{odd} or continue along the cross aisle in order to skip the next two aisles.

The decision points are numbered ascending from left to right. We will derive formulas to calculate $q_{j,i}$ at these decision points based on the following approach. First, for a given number of picks per order, n, and a given number of picking segments in the order picking system, νh, we can basically calculate the number of different possibilities to distribute the picks to the segments. From this number we can then select those cases which would force a picker to visit a certain decision point. Finally we can select those cases which force the picker to be at a certain decision point *and* require a certain movement beyond that decision point. From the last two numbers we can calculate the probability of a specific movement.

Calculating the number of different possibilities to distribute n picks to νh segments can be done by using a very simple urn model (Johnson and Kotz 1977, p. 2). Suppose we have an urn containing a total of νh balls of which n balls are black and $w = \nu h - n$ balls are white. A black ball indicates a *pick*, a white ball indicates *no pick*. We chose without replacement and the color of the first ball drawn defines if there is a pick at segment 1, the color of the second ball drawn defines if there is a pick at segment 2 etcetera. We find the number of distinguishable draws as:

$$\binom{\nu h}{n} = \frac{\nu h!}{n!(\nu h - n)!} \tag{4.2}$$

This is exactly the number of different possibilities to distribute n picks among the νh segments.

4.5.3 Transition Probabilities at Upper Cross Aisle Decision Points A_x

In an order picking system with ν aisles we have $\left(\frac{\nu}{2} - 1\right)$ decision points of type A. The decision points are represented by an elementary queueing system which in turn represents the segment of the upper cross aisle which is located directly above an even aisle. Due to the routing policy, the pickers will only pass through decision point A_x for a subset of all possible allocations of picks.

In order to simplify the resulting formula to calculate $q_{j,i}$, we number the decision points with indices $x = 0, 1, 2, 3, ..., (\frac{\nu}{2} - 2)$. Decision point A_0 immediately precedes aisle 2, decision point A_1 immediately precedes aisle 4 and generally decision point A_x immediately precedes the even aisle $\alpha_{even} = 2x + 2$ and thus $x = \frac{\alpha_{even}-2}{2}$.

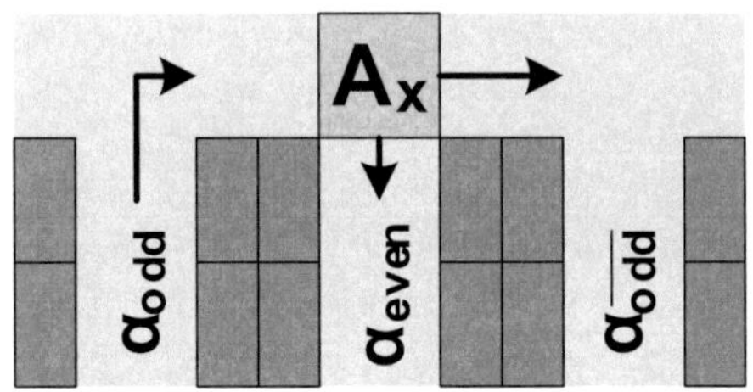

Figure 4.7: Notations at decision point A_x

We can record that A_x is visited whenever:

- at least one pick is in aisle α_{even} or in aisle $\alpha_{odd} = \alpha_{even} - 1$

- no pick is made in aisles α_{even} and α_{odd} and at least one pick is made in any odd aisle β_{odd} with $\beta_{odd} < \alpha_{odd}$. At the same time no pick is made in any aisle β with $\beta_{odd} < \beta < \alpha_{odd}$ and at least one pick is made in any aisle $\alpha > \alpha_{even}$. Furthermore, the number of picks in the aisles $\alpha < \beta_{odd}$ must not exceed $(n-2)$ and the number of picks in aisles $\alpha \le \beta_{odd}$ must be smaller than n.

Bullet point 1 states that A_x will be used if there are picks in the aisles directly neighboring the decision point. Bullet point 2 is more complex. It represents cases where A_x will only be passed in order to skip α_{odd}, α_{even} and $\alpha_{\overline{odd}}$. This in turn will only happen if the last pick in aisles smaller than α_{odd} is done in an odd aisle and at least one pick is in any aisle bigger than α_{even}. The odd aisle β_{odd} will carry the picker to the upper cross aisle. The picker stays on that upper cross aisle until reaching A_x as long as no even aisle will carry him back to the lower cross aisle. The number of picks in aisles $\alpha \leq \beta_{odd}$ must be smaller than n because for n the picker would return to the depot and never reach A_x. Because at least one pick is done in aisle β_{odd} and at least one pick is done in $\alpha > \alpha_{even}$, there should be at most $(n-2)$ picks in aisles $\alpha < \beta_{odd}$. Otherwise the picks in these aisles would combine to n and the picker would return to the depot.

For further clarification, we discuss a small example, reconsidering figure 4.6. Let the number of picks be $n = 4$. We will focus on decision point A_2 which is located directly above aisle 6. Obviously there will a visit to A_2 if picks are located in aisles 5 or 6 (bullet point 1). If no picks are in aisles 5 or 6, A_2 will still be visited to skip the aisles (bullet point 2) in the following cases:

- No picks in aisles 1 and 2: one pick in aisle 3 and three picks in aisles 7 to 10 or two picks in aisle 3 and two picks in aisles 7 to 10 or three picks in aisle 3 and one pick in aisles 7 to 10.

- One pick in aisles 1 and 2: one pick in aisle 3 and two picks in aisles 7 to 10 or two picks in aisle 3 and one pick in aisles 7 to 10

- Two picks in aisles 1 and 2: one pick in aisle 3 and one pick in aisles 7 to 10.

We will now present a general approach to calculate the transition probability at decision point A_x. We define Γ_{A_x} as the total number of cases for which A_x is visited. Additionally, we define $\Gamma_{A_x|\alpha_{next}\neq\alpha_{even}}$ as the number of cases for which decision point A_x is visited and aisle α_{even} is not visited. We can then calculate the transition probability, i.e. the probability that aisle α_{even} is the next aisle entered by:

$$q_{A_x,\alpha_{even}} = 1 - \frac{\Gamma_{A_x|\alpha_{next}\neq\alpha_{even}}}{\Gamma_{A_x}}$$

First, we calculate the number of cases with at least one pick in aisles α_{odd} or α_{even} as follows. We have to allocate y picks to α_{odd} and α_{even}, where $y = (1, 2, ...n)$. With equation (4.2) this is

$$\binom{2h}{y}$$

The binomial coefficient $\binom{n}{k}$ equals 0 for $n < k$ (Bronstein and Semendjajew 1991, p. 104). This reflects the assumption that at most h picks can be located in one aisle. The remaining $(n - y)$ picks are allocated to the remaining $(\nu h - 2h)$ segments:

$$\binom{\nu h - 2h}{n - y}$$

Combining all possibilities to split the picks on the aisles, we get:

$$\sum_{y=1}^{n} \binom{2h}{y} \binom{\nu h - 2h}{n - y} \tag{4.3}$$

Secondly, we also consider the cases, for which decision point A_x is visited even though picks are neither located in α_{odd} nor in α_{even}. Let z be the number of picks in β_{odd} and y be the picks in all aisles $\alpha < \beta_{odd}$. The auxiliary variable o includes the information which odd aisle β_{odd} was used to travel to the upper cross aisle. For $o = 1$ the first odd aisle was used, for $o = 2$ the second odd aisle was used and so forth. We can use o to count for the number of aisles preceding β_{odd} and distribute y picks to these aisles $\alpha < \beta_{odd}$ by:

$$\binom{2(o - 1)h}{y}$$

In aisle β_{odd} we can distribute z picks according to:

$$\binom{h}{z}$$

The remaining $(n - y - z)$ picks have to be distributed on all aisles $\alpha > \alpha_{even}$:

$$\binom{h(\nu - (2x + 2))}{n - y - z}$$

As discussed earlier the number of picks y must not exceed $(n - 2)$ and the combined number of picks $(y + z)$ must be smaller than n.

By combining all possibilities we get:

$$\sum_{y=0}^{n-2}\sum_{z=1}^{n-1-y}\sum_{o=1}^{x}\binom{2(o-1)h}{y}\binom{h}{z}\binom{h(\nu-(2x+2))}{n-y-z} \tag{4.4}$$

Overall, the number of cases Γ_{A_x} for which decision point A_x is visited can be calculated by adding up formulas (4.3) and (4.4):

$$\Gamma_{A_x} = \sum_{y=1}^{n}\binom{2h}{y}\binom{\nu h - 2h}{n-y}$$

$$+\sum_{y=0}^{n-2}\sum_{z=1}^{n-1-y}\sum_{o=1}^{x}\binom{2(o-1)h}{y}\binom{h}{z}\binom{h(\nu-(2x+2))}{n-y-z} \tag{4.5}$$

for all $x = 1, 2, 3, ..., (\frac{\nu}{2} - 2)$. For $x = 0$ we only consider formula (4.3) as A_0 is never used for skipping aisles:

$$\Gamma_{A_0} = \sum_{y=1}^{n}\binom{2h}{y}\binom{\nu h - 2h}{n-y} \tag{4.6}$$

Next, we need to consider the cases for which aisle α_{even} is not entered. This equals a situation in which a picker is standing at decision point A_x with picks remaining but with no picks in aisles α_{even} or the succeeding aisle $(\alpha_{even} + 1) = \alpha_{\overline{odd}} = (2x+3)$. α_{even} will not be entered if there is a pick in an odd aisle $\beta_{odd} \leq \alpha_{odd}$, while at the same time no pick is made in any aisle β with $\beta_{odd} < \beta < \alpha_{odd}$ and at least one pick is made in any aisle $\alpha > \alpha_{\overline{odd}}$. The number of picks in aisles $\alpha \leq \beta_{odd}$ must be smaller than n because for n the picker would return to the depot. Because at least one pick is done in aisle β_{odd} and at least one pick is done in $\alpha > \alpha_{\overline{odd}}$, there should be at most $(n-2)$ picks in aisles $\alpha < \beta_{odd}$.

Like before, let y be the picks in all aisles $\alpha < \beta_{odd}$ and z be the picks in β_{odd} and o be an auxiliary variable to count for aisles preceding β_{odd}. We distribute y picks to aisles $\alpha < \beta_{odd}$ in

$$\binom{2oh}{y}$$

different ways. Simultaneously, we distribute z picks to β_{odd} in

$$\binom{h}{z}$$

different ways. The remaining $(n - y - z)$ picks are distributed to the aisles $\alpha > \alpha_{\overline{odd}}$ according to:

$$\binom{h(\nu - (2x + 3))}{n - y - z}$$

Combining these three terms yields the number of cases for which α_{even} is not entered at decision point A_x:

$$\Gamma_{A_x | \alpha_{next} \neq \alpha_{even}} = \sum_{y=0}^{n-2} \sum_{z=1}^{n-1-y} \sum_{o=0}^{x} \binom{2oh}{y} \binom{h}{z} \binom{h(\nu - (2x + 3))}{n - y - z} \tag{4.7}$$

$$\Gamma_{A_0 | \alpha_{next} \neq 2} = \sum_{y=0}^{n-2} \sum_{z=1}^{n-1-y} \binom{h}{z} \binom{h(\nu - 3)}{n - y - z} \tag{4.8}$$

With this we can finally calculate the transition probability at decision point A_x:

$$q_{A_x, \alpha_{even}} = 1 - \frac{\Gamma_{A_x | \alpha_{next} \neq \alpha_{even}}}{\Gamma_{A_x}} \tag{4.9}$$

$$q_{A_0, \alpha=2} = 1 - \frac{\Gamma_{A_0 | \alpha_{next} \neq 2}}{\Gamma_{A_0}} \tag{4.10}$$

4.5.4 Transition Probabilities at within-aisle Decision Points B_x

Decision points of type B can be found at the very last location of an even aisle α_{even}, which is the location closest to the lower cross aisle. For ν aisles, we thus have $(\frac{\nu}{2} - 1)$ decision points of type B. The picker has to decide whether to continue his picking tour or to return to the depot. The latter is obviously the case if all n picks have been picked in aisles $\alpha \leq \alpha_{even}$. Again we number the decision points with indices $x = 0, 1, 2, 3, ..., (\frac{\nu}{2} - 2)$. With this B_0 is the last location of aisle 2, B_1 is the last location of aisle 4 and generally B_x is the last location of aisle $\alpha_{even} = (2x + 2)$ thus $x = \frac{\alpha_{even}-2}{2}$.

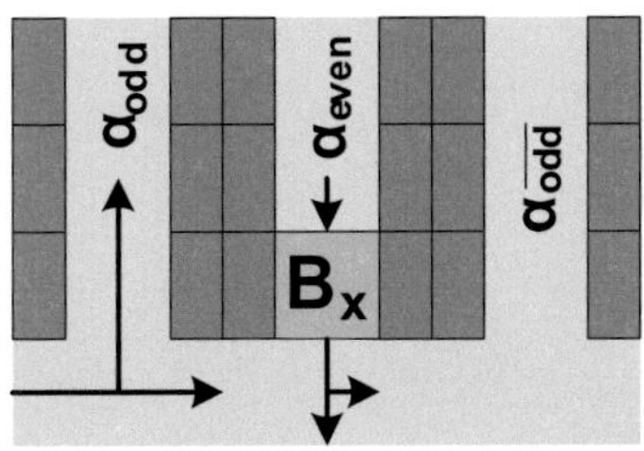

Figure 4.8: Notations at decision point B_x

We visit decision point B_x whenever:

- at least one pick is done in aisle α_{even}

- no pick is made in aisle α_{even} and at least one pick is done in aisle α_{odd} and all other picks are done in aisles $\alpha < \alpha_{odd}$

- no pick is made in α_{even} and at least one pick is made in any odd aisle $\beta_{odd} < \alpha_{even}$, while at the same time no pick is made in any aisle β with $\beta_{odd} < \beta < \alpha_{even}$ and at least one pick is made in the odd aisle $\alpha_{\overline{odd}} = \alpha_{even} + 1$

In order to calculate the number of cases Γ_{B_x} for which decision point B_x is visited we can use some of the formulas we derived for decision points of type A. Γ_{A_x} returns the number of different possibilities to be at decision point A_x and $\Gamma_{A_x|\alpha_{next}\neq\alpha_{even}}$ yields the number of cases to be at decision point A_x and not enter the even aisle, but continue along the cross aisle instead. Consequently the number of possibilities to enter the even aisle and hence Γ_{B_x} can be calculated by:

$$\Gamma_{B_x} = \Gamma_{A_x} - \Gamma_{A_x|\alpha_{next}\neq\alpha_{even}} \tag{4.11}$$

In order to obtain a transition probability, we have to know $\Gamma_{B_x|\alpha_{next}=\emptyset}$, i.e. the number of cases for which decision B_x is visited *and* the picker continues towards the return path. For this case, we have to distribute n picks to all aisles $\alpha \leq \alpha_{even} = (2x+2)$ with at least one pick in aisles α_{odd} or α_{even}. We can have y picks in these two aisles in

$$\binom{2h}{y}$$

different ways. We can distribute the remaining $(n-y)$ picks to all aisles preceding α_{odd} in

$$\binom{2xh}{n-y}$$

different ways. $\Gamma_{B_x|\alpha_{next}=\emptyset}$ is obtained by considering all options to split the n picks on the aisles:

$$\Gamma_{B_x|\alpha_{next}=\emptyset} = \sum_{y=1}^{n} \binom{2h}{y}\binom{2xh}{n-y} \tag{4.12}$$

$$\Gamma_{B_0|\alpha_{next}=\emptyset} = \sum_{y=1}^{n} \binom{2h}{y} \tag{4.13}$$

The transition probability at decision point B_x is then given by:

$$q_{B_x, ReturnPath} = \frac{\Gamma_{B_x|\alpha_{next}=\emptyset}}{\Gamma_{B_x}} \tag{4.14}$$

$$q_{B_0, ReturnPath} = \frac{\Gamma_{B_0|\alpha_{next}=\emptyset}}{\Gamma_{A_0} - \Gamma_{A_0|\alpha_{next}\neq 2}} \tag{4.15}$$

4.5.5 Transition Probabilities at Lower Cross Aisle Decision Points C_x

For ν aisles an order picking system has $\left(\frac{\nu}{2}-1\right)$ decision points of type C. The decision points are located on the lower cross aisle segments immediately in front of an odd aisle α_{odd}. The decision points are numbered with indices $x = 1, 2, 3, ..., (\frac{\nu}{2} - 2)$ thus decision point C_1 is located directly in front of aisle 3, decision point C_2 is located directly in front of aisle 5 and generally decision point C_x is located immediately in front of the odd aisle $\alpha_{odd} = 2x + 1$ thus $x = \frac{\alpha_{odd} - 1}{2}$. Decision point C_0 is located right after the depot and is visited for every order.

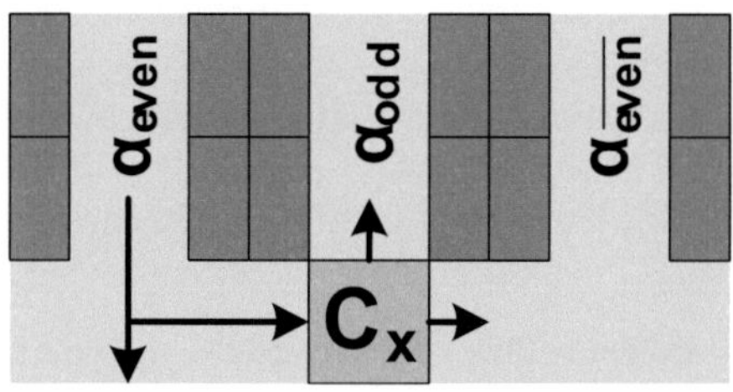

Figure 4.9: Notations at decision point C_x

Like for the other decision points, we need to determine the number of possible pick allocations leading to a visit of C_x. This will happen whenever:

- all picks are located in aisles $\alpha > \alpha_{even}$

- no pick is made in aisle α_{odd} and at least one pick is made in any even aisle $\beta_{even} < \alpha_{odd}$, while at the same time no pick is made in any aisle β with $\beta_{even} < \beta < \alpha_{odd}$ and at least one pick is made in any aisle $\alpha > \alpha_{odd}$. Furthermore the number of picks in the aisles $\alpha < \beta_{even}$ must not exceed $(n-2)$ and the number of picks in aisles $\alpha \leq \beta_{even}$ must be smaller than n.

- at least one pick is done in aisle α_{odd} and at least one pick is done in aisles $\alpha < \alpha_{odd}$ and there might be picks in aisles $\alpha > \alpha_{odd}$.

The first case can be formulated straightaway. We have to distribute all n picks to the aisles $\alpha > \alpha_{even}$ and this can be done in

$$\binom{h(\nu - 2x)}{n} \tag{4.16}$$

different ways.

For the second case, let z be the number of picks in aisle β_{even} and y be the number of picks in aisles $\alpha < \beta_{even}$. Again we use the auxiliary variable o to count for the number of aisles preceding β_{even}. We can then distribute y picks to these aisles $\alpha < \beta_{even}$ in

$$\binom{(2o-1)h}{y}$$

different ways. At the same time we find z picks in aisle β_{even} in

$$\binom{h}{z}$$

different ways. As we have no picks for aisle α_{odd} we have to allocate the remaining $(n-y-z)$ picks to the remaining aisles $\alpha > \alpha_{odd}$ and can do this in in

$$\binom{h(\nu-(2x+1))}{n-y-z}$$

different ways. We can combine the three terms of case 2 and get:

$$\sum_{y=0}^{n-2}\sum_{z=1}^{n-1-y}\sum_{o=1}^{x}\binom{(2o-1)h}{y}\binom{h}{z}\binom{h(\nu-(2x+1))}{n-y-z} \tag{4.17}$$

In the last case we have at least one pick in aisles $\alpha < \alpha_{odd}$ and we can distribute the z picks in

$$\binom{2xh}{z}$$

different ways. At the same time we have y picks in the aisles $\alpha > \alpha_{odd}$ and depending on x and there are

$$\binom{h(\nu-(2x+1))}{y}$$

different possibilities to do so. Finally the remaining $(n-y-z)$ picks are located in aisle α_{odd} and for this we find

$$\binom{h}{n-y-z}$$

different possibilities. The three terms of case 3 then combine to:

$$\sum_{y=0}^{n-2}\sum_{z=1}^{n-1-y}\binom{2xh}{z}\binom{h(\nu-(2x+1))}{y}\binom{h}{n-y-z}$$ (4.18)

We combine formulas (4.16), (4.17) and (4.18) to obtain the number of cases Γ_{C_x} for which decision point C_x is visited:

$$\Gamma_{C_x} = \binom{h(\nu-2x)}{n}$$

$$+\sum_{y=0}^{n-2}\sum_{z=1}^{n-1-y}\sum_{o=1}^{x}\binom{(2o-1)h}{y}\binom{h}{z}\binom{h(\nu-(2x+1))}{n-y-z}$$

$$+\sum_{y=0}^{n-2}\sum_{z=1}^{n-1-y}\binom{2xh}{z}\binom{h(\nu-(2x+1))}{y}\binom{h}{n-y-z}$$ (4.19)

for all $x = 1, 2, 3, ..., (\frac{\nu}{2} - 2)$. Decision point C_0 is always visited, i.e. for all possible allocations of picks:

$$\Gamma_{C_0} = \binom{h\nu}{n}$$ (4.20)

The reader might wonder why formula (4.19) is slightly more complicated than formula (4.5) for decision points A even though decision points A and C seem to be logically related to each other. For decision point A we were able to assume that it is visited either when there is at least one pick in the corresponding aisles or in the aisles $\alpha < \alpha_{even}$. This assumption does not hold for decision point C as there might be cases when the picker has finished his order after aisle α_{even}. Such cases are considered by decision point B though and in order to avoid double counts, the procedure for decision point C is differed, resulting in a formula slightly more complicated.

Building on formula (4.19), we can derive a transition probability by considering those cases for which the aisle α_{odd} is not entered, thus skipping aisles α_{odd} and $\alpha_{\overline{even}}$. This occurs for the following two cases:

- all picks are done in aisles $\alpha > \alpha_{\overline{even}}$
- no picks are made in aisles α_{odd} or $\alpha_{\overline{even}}$ and at least one pick is made in any even aisle $\beta_{even} < \alpha_{odd}$, while at the same time no pick is made in any aisle β with $\beta_{even} < \beta < \alpha_{odd}$ and at least one pick is made in any aisle $\alpha > \alpha_{\overline{even}}$. Furthermore the number of picks in the aisles $\alpha < \beta_{even}$ must not exceed $(n-2)$ and the number of picks in aisles $\alpha \leq \beta_{even}$ must be smaller than n.

The first case can be quantified by calculating

$$\binom{h(\nu - (2x+2))}{n} \tag{4.21}$$

The second case is derived according to previous considerations. Let z be the number of picks in aisle β_{even}, let y be the picks in all aisles $\alpha < \beta_{even}$ and let $(n - y - z)$ be the number of picks in aisles $\alpha > \alpha_{\overline{even}}$. Then we can identify

$$\sum_{y=0}^{n-2} \sum_{z=1}^{n-1-y} \sum_{o=1}^{x} \binom{(2o-1)h}{y} \binom{h}{z} \binom{h(\nu - (2x+2))}{n - y - z} \tag{4.22}$$

different possibilities to allocate the picks.

Combining formulas (4.21) and (4.22) results in the overall number of cases $\Gamma_{C_x|\alpha_{next}\neq\alpha_{odd}}$, for which decision point C_x is visited *and* the picking route is continuing along the lower cross aisle:

$$\Gamma_{C_x|\alpha_{next}\neq\alpha_{odd}} = \binom{h(\nu - (2x+2))}{n}$$

$$+ \sum_{y=0}^{n-2} \sum_{z=1}^{n-1-y} \sum_{o=1}^{x} \binom{(2o-1)h}{y} \binom{h}{z} \binom{h(\nu - (2x+2))}{n - y - z} \tag{4.23}$$

For $x = 0$, we only consider formula (4.21) as there can not be picks in aisles $\alpha < \alpha_{odd}$:

$$\Gamma_{C_0|\alpha_{next}\neq 1} = \binom{h(\nu - 2)}{n} \tag{4.24}$$

We calculate the transition probability at decision point C_x according to:

$$q_{C_x,\alpha_{odd}} = 1 - \frac{\Gamma_{C_x|\alpha_{next}\neq\alpha_{odd}}}{\Gamma_{C_x}} \qquad (4.25)$$

$$q_{C_0,\alpha=1} = 1 - \frac{\Gamma_{C_0|\alpha_{next}\neq 1}}{\Gamma_{C_0}} \qquad (4.26)$$

4.6 Modeling of One-Way Traversal Routing for within-aisle Class-Based Storage Policy

Besides the random storage policy, we will also analyze systems in which items are stored according to their picking frequency. The transition probabilities $q_{j,i}$ and thus visit ratios e_i should change accordingly because some parts of the system will be visited more often than others. We will consider a class-based storage policy with L^{CB} classes as described in chapter 2.3.1. The storage policy will be within-aisle, i.e. the most popular class will be in the leftmost aisles, i.e. the aisles closest to the depot and the least popular class will be in the rightmost aisles, i.e. the aisles farthest from the depot.

4.6.1 Extensions of the Random Storage Policy Models

As the structure of the order picking system remains unchanged, so do predefined movements for class-based storage. For non-predefined movements at the decision points, we will again use the approach of counting cases which result in a certain movement of the picker. Again we apply the urn model of Johnson and Kotz (1977, p. 2) as described on page 71. In contrast to random storage policy we now have to incorporate the fact that some segments have a higher probability of being selected. We achieve this by introducing the concept of virtual aisle segments.

Introduction of Virtual Aisle Segments

Again we consider an urn model containing n black balls and w white balls with $n + w = \nu h$. We choose without replacement and the draw determines if a segment is selected for picking (black balls) or non-picking (white balls). In order to increase the probability to allocate a black ball to segment i we create j virtual segments and j extra white balls. We treat the event of allocating a black ball to any of the j virtual segments as if the black ball was allocated to segment i.

Figure 4.10 illustrates an example of this logic. For two aisles with two segments each and one pick, the probability that the pick will be in aisle 1 equals $\frac{1}{2}$. After we add two virtual segments and two white balls (non-picks) to the urn and treat the virtual segments as if they belonged to aisle 1, the probability will increase to $\frac{2}{3}$. Thus we have *virtually* increased the size of aisle 1 by introducing new *virtual aisle segments*.

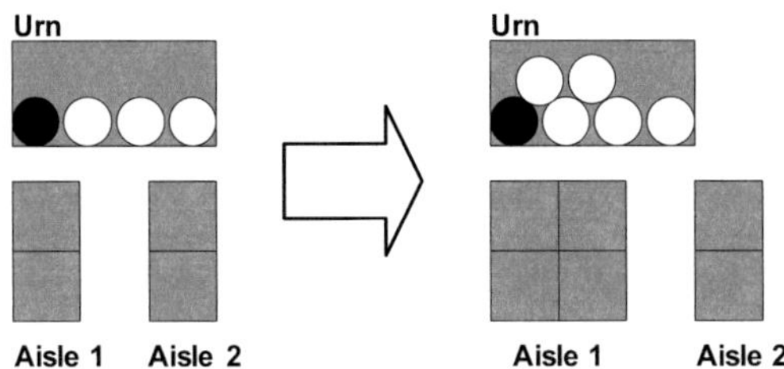

Figure 4.10: Concept of virtual aisle segments - example

If we apply this concept to aisle α, the number of segments is virtually increased from h to h_α, i.e. h_α is the number of segments in the virtual aisle. We now present a general approach to derive this number as a function of the size and picking frequencies of the respective classes and the original number of segments per aisle h.

Let $\gamma_{p,seg}$ be the percentage of segments belonging to class γ. Assuming that each segment is assigned to only one class, we can calculate the absolute number of segments γ_{seg} belonging to class γ by:

$$\gamma_{seg} = \lceil \gamma_{p,seg} \cdot \nu \cdot h \rceil \tag{4.27}$$

For the last class L^{CB} we have:

$$L_{seg}^{CB} = \nu \cdot h - \sum_{\forall \gamma \neq L^{CB}} \gamma_{seg} \tag{4.28}$$

Let $\gamma_{p,picks}$ be the percentage of picks done in class γ. We introduce an auxiliary variable $\gamma_{virtual}$, which represents a multiplying factor used to increase the aisle to its virtual size. As we do not convert the number of segments in the last class we define

$$L_{virtual}^{CB} = 1$$

Using formulas (4.27) and (4.28), we can calculate $\gamma_{virtual}$ for all remaining classes:

$$\gamma_{virtual} = \frac{L_{seg}^{CB} \cdot \frac{\gamma_{p,picks}}{L_{p,picks}^{CB}}}{\gamma_{seg}} \tag{4.29}$$

The term $\frac{\gamma_{p,picks}}{L_{p,picks}^{CB}}$ indicates how much more likely a pick from class γ is in comparison to a pick from class L^{CB}. We obtain the number of segments that will cover class γ by multiplying this with L_{seg}^{CB}. Division by γ_{seg} will result in the multiplying factor $\gamma_{virtual}$.

We can use $\gamma_{virtual}$ to calculate the number of segments in the virtual aisle h_i. Starting with the leftmost aisle, we determine how many of the segments of that aisle belong to a certain class. There are three different cases:

- Case 1: the number of segments in the A-class equals the number of segments in that aisle, i.e. $A_{seg} = h$. We can then calculate h_i by:

$$h_i = A_{virtual} \cdot h$$

- Case 2: the number of segments in the A-class is smaller than the number of segments in that aisle, i.e. $A_{seg} < h$. We will immediately share the aisle with the next class and calculate h_i according to:

$$h_i = A_{virtual} \cdot A_{seg} + B_{virtual} \cdot (h - A_{seg})$$

- Case 3: the number of segments in the A-class is bigger than the number of segments in that aisle, i.e. $A_{seg} > h$. We then have to distribute the class segments to several aisles. $\lfloor \frac{A_{seg}}{h} \rfloor$ aisles will be exclusively of class A. The remaining locations $A_{seg,rem} = (A_{seg} - \lfloor \frac{A_{seg}}{h} \rfloor \cdot h)$ will share an aisle with segments of the B-class. For that shared aisle h_i will be calculated by:

$$h_i = A_{virtual} \cdot A_{seg,rem} + B_{virtual} \cdot (h - A_{seg,rem})$$

- Note that in cases 2 and 3 we implicitly assumed that the shared aisle holds at most two different classes. It can occur though that B_{seg} will not be enough to fill the shared aisle so we have to start adding segments of another class. We simply have to repeat the procedures of cases 2 and 3 until the aisle is completely filled. For example, three clases in an aisle will result in:

$$h_i = A_{virtual} \cdot A_{seg} + B_{virtual} \cdot B_{seg} + C_{virtual} \cdot (h - A_{seg} - B_{seg})$$

Generally, for j classes in an aisle, we get:

$$h_i = \sum_{\forall k \leq (j-1)} k_{virtual} \cdot k_{seg} + j_{virtual} \cdot (h - \sum_{\forall k \leq (j-1)} k_{seg})$$

We repeat this procedure until all aisles have been extended from their original size h to their virtual size h_i.

Example for Virtual Aisle Segments

We present an example to illustrate the procedure by considering an order picking system with $\nu = 6$ and $h = 4$. Three classes with the following characteristics are introduced:

$$A_{p,picks} = 60\% \quad A_{p,seg} = 16.\bar{6}\%$$

$$B_{p,picks} = 30\% \quad B_{p,seg} = 33.\bar{3}\%$$

$$C_{p,picks} = L^{CB}_{p,picks} = 10\% \quad C_{p,seg} = L^{CB}_{p,seg} = 50\%$$

With (4.27) and (4.28) we get:

$$A_{seg} = \lceil \frac{1}{6} \cdot 6 \cdot 4 \rceil = 4 \qquad B_{seg} = \lceil \frac{1}{3} \cdot 6 \cdot 4 \rceil = 8 \qquad C_{seg} = 24 - 4 - 8 = 12$$

We calculate $A_{virtual}$ according to:

$$A_{virtual} = \frac{L^{CB}_{seg} \cdot \frac{A_{p,picks}}{L^{CB}_{p,picks}}}{A_{seg}} = \frac{12 \cdot \frac{0.6}{0.1}}{4} = 18$$

In the same way we obtain $B_{virtual}$:

$$B_{virtual} = \frac{L^{CB}_{seg} \cdot \frac{B_{p,picks}}{L^{CB}_{p,picks}}}{B_{seg}} = \frac{12 \cdot \frac{0.3}{0.1}}{8} = 4.5$$

As the C-class is the last class we define:

$$C_{virtual} = 1$$

We can now calculate the length of the virtual aisles. Starting with aisle 1, we find that $A_{seg} = 4 = h$ (Case 1) leading to:

$$h_1 = A_{virtual} \cdot h = 18 \cdot 4 = 72$$

For aisle 2 we find that $B_{seg} = 8 > 4 = h$ (Case 3) and consequently distribute the class on $\lfloor \frac{B_{seg}}{h} \rfloor = \frac{8}{4} = 2$ aisles with $B_{seg,rem} = 0$. Hence we have:

$$h_2 = B_{virtual} \cdot h = 4.5 \cdot 4 = 18$$

$$h_3 = B_{virtual} \cdot h = 4.5 \cdot 4 = 18$$

For the final class we use $C_{virtual} = 1$ and get:

$$h_4 = h_5 = h_6 = C_{virtual} \cdot h = 1 \cdot 4 = 4$$

Overall we have

$$\sum_{i=1}^{\nu} h_i = 72 + 18 + 18 + 4 + 4 + 4 = 120$$

For $n = 1$ the probability to have that pick in aisle 1 is

$$\frac{\binom{72}{1}}{\binom{120}{1}} = \frac{72}{120} = 0.6$$

which equals the original $A_{p,picks} = 60\%$. For $n = 2$ the probability to have two picks in aisle 1 is

$$\frac{\binom{72}{2}}{\binom{120}{2}} = \frac{2556}{7140} = 0,3579 < 0,36 = 0,6 \cdot 0,6$$

This result confirms a basic assumption stemming from of our original urn model. If the first ball selected has a 60% chance to be black then the probability of the second ball being black will be slightly lower than 60% as we choose without replacement.

We should note that the above formulas for $\gamma_{virtual}$ and h_i might not always result in integer values. However, the binomial coefficient is defined for integer values only. Therefore we need to either round up or round down. This step can result in minor modeling inaccuracies. Experiments have shown that the accuracy improvement is negligible, thus we hold on to the presented procedure.

In order to calculate the transition probabilities at decision points A_x, B_x and C_x we will again use the approach of counting the number of pick allocations which will lead to a visit of a certain decision point (see chapter 4.5). From this number we will then select those cases which result in a certain movement of the picker. The urn model of Johnson and Kotz (1977, p. 2) is combined with the virtual aisle segments h_i to consider the fact that some segments have a higher probability of being selected. Basically we reuse the formulas of chapter 4.5 as the logic of allocating picks remains unchanged. We just incorporate h_i in all terms where h was used.

4.6.2 Transition Probabilities at Upper Cross Aisle Decision Points A_x

We use the cross aisle at decision point A_x whenever we have at least one pick in the adjacent aisles α_{odd} or α_{even} or if we skip those aisles after having used an odd aisle $\beta_{odd} < \alpha_{odd}$. α_{odd} will carry index $(2x+1)$ and α_{even} will be denoted by index $(2x+2)$. For the first case we allocate y picks to α_{odd} and α_{even}:

$$\sum_{y=1}^{n} \binom{h_{2x+1} + h_{2x+2}}{y} \qquad \text{s.t. } y \leq 2h$$

The remaining $(n-y)$ picks will be allocated to all other aisles:

$$\binom{\sum_{i=1}^{\nu} h_i - h_{2x+1} - h_{2x+2}}{n-y} \qquad \text{s.t. } n-y \leq h(\nu-2)$$

We consider all possible combinations according to:

$$\sum_{y=1}^{n} \binom{h_{2x+1} + h_{2x+2}}{y} \binom{\sum_{i=1}^{\nu} h_i - h_{2x+1} - h_{2x+2}}{n-y} \tag{4.30}$$

For the second case (skipping α_{odd} and α_{even}) we reuse formula (4.4) and get:

$$\sum_{y=0}^{n-2} \sum_{z=1}^{n-1-y} \sum_{o=1}^{x} \binom{\sum_{i=1}^{2(o-1)} h_i}{y} \binom{h_{2o-1}}{z} \binom{\sum_{i=2x+3}^{\nu} h_i}{n-y-z} \tag{4.31}$$

There are three necessary conditions for formula (4.31):

$$y \leq h(2o-2) \qquad z \leq h \qquad (n-y-z) \leq h(\nu-(2x+2))$$

We add up formulas (4.30) and (4.31) to calculate the number of cases $\Gamma_{A_x}^{CB}$ for which decision point A_x is visited:

$$\Gamma_{A_x}^{CB} = \sum_{y=1}^{n} \binom{h_{2x+1} + h_{2x+2}}{y} \binom{\sum_{i=1}^{\nu} h_i - h_{2x+1} - h_{2x+2}}{n-y}$$

$$+ \sum_{y=0}^{n-2} \sum_{z=1}^{n-1-y} \sum_{o=1}^{x} \binom{\sum_{i=1}^{2(o-1)} h_i}{y} \binom{h_{2o-1}}{z} \binom{\sum_{i=2x+3}^{\nu} h_i}{n-y-z} \tag{4.32}$$

For $x = 0$ we only consider formula (4.30) as A_0 is never used for skipping aisles and get:

$$\Gamma_{A_0}^{CB} = \sum_{y=1}^{n} \binom{h_{2x+1} + h_{2x+2}}{y} \binom{\sum_{i=1}^{\nu} h_i - h_{2x+1} - h_{2x+2}}{n - y} \qquad (4.33)$$

Next we consider the cases $\Gamma_{A_x}^{CB}$ for which A_x is visited but aisle α_{even} is not entered. We reuse formula (4.7) and calculate the number of cases $\Gamma_{A_x|\alpha_{next}\neq\alpha_{even}}^{CB}$ as follows:

$$\Gamma_{A_x|\alpha_{next}\neq\alpha_{even}}^{CB}$$
$$= \sum_{y=0}^{n-2} \sum_{z=1}^{n-1-y} \sum_{o=0}^{x} \binom{\sum_{i=1}^{2o} h_i}{y} \binom{h_{2o+1}}{z} \binom{\sum_{i=2x+4}^{\nu} h_i}{n - y - z} \qquad (4.34)$$

$$\Gamma_{A_0|\alpha_{next}\neq 2}^{CB} = \sum_{y=0}^{n-2} \sum_{z=1}^{n-1-y} \binom{h_1}{z} \binom{\sum_{i=4}^{\nu} h_i}{n - y - z} \qquad (4.35)$$

Formula (4.34) is valid if the following conditions hold:

$$y \leq h(2o) \qquad z \leq h \qquad (n - y - z) \leq h(\nu - (2x + 3))$$

With this we calculate the transition probabilities at decision point A_x by:

$$q_{A_x,\alpha_{even}}^{CB} = 1 - \frac{\Gamma_{A_x|\alpha_{next}\neq\alpha_{even}}^{CB}}{\Gamma_{A_x}^{CB}} \qquad (4.36)$$

$$q_{A_0,\alpha=2}^{CB} = 1 - \frac{\Gamma_{A_0|\alpha_{next}\neq 2}^{CB}}{\Gamma_{A_0}^{CB}} \qquad (4.37)$$

4.6.3 Transition Probabilities at within-aisle Decision Points B_x

Again we can use the approach presented for random storage. We subtract formula (4.34) from formula (4.32) to calculate the total number of cases $\Gamma^{CB}_{B_x}$ for which decision point B_x will be visited:

$$\Gamma^{CB}_{B_x} = \Gamma^{CB}_{A_x} - \Gamma^{CB}_{A_x|\alpha_{next}\neq\alpha_{even}} \tag{4.38}$$

$$\Gamma^{CB}_{B_0} = \Gamma^{CB}_{A_0} - \Gamma^{CB}_{A_0|\alpha_{next}\neq\alpha_{even}} \tag{4.39}$$

We can easily derive the number of cases $\Gamma^{CB}_{B_x|\alpha_{next}=\emptyset}$ for which B_x is visited *and* all n picks have been done, i.e. the picker returning to the depot. This is the case if we have at least one pick in aisle α_{even} or aisle $\alpha_{odd} = (\alpha_{even} - 1)$ while the remaining $(n-y)$ picks are located in aisles $\alpha < \alpha_{odd}$. Again, α_{odd} will carry index $(2x+1)$, α_{even} will be denoted by index $(2x+2)$. The number of cases is given by:

$$\Gamma^{CB}_{B_x|\alpha_{next}=\emptyset} = \sum_{y=1}^{n} \binom{h_{2x+1} + h_{2x+2}}{y} \binom{\sum_{i=1}^{2x} h_i}{n-y} \tag{4.40}$$

$$\Gamma^{CB}_{B_0|\alpha_{next}=\emptyset} = \sum_{y=1}^{n} \binom{h_1 + h_2}{y} \tag{4.41}$$

For (4.40) we have the following conditions:

$$y \leq 2h \qquad (n-y) \leq 2xh$$

The transition probability at decision point B_x is then given by:

$$q^{CB}_{B_x,ReturnPath} = \frac{\Gamma^{CB}_{B_x|\alpha_{next}=\emptyset}}{\Gamma^{CB}_{A_x} - \Gamma^{CB}_{A_x|\alpha_{next}\neq\alpha_{even}}} \tag{4.42}$$

$$q^{CB}_{B_0,ReturnPath} = \frac{\Gamma^{CB}_{B_0|\alpha_{next}=\emptyset}}{\Gamma^{CB}_{A_0} - \Gamma^{CB}_{A_0|\alpha_{next}\neq 2}} \tag{4.43}$$

4.6.4 Transition Probabilities at Lower Cross Aisle Decision Points C_x

The number of cases for which decision point C_x is visited can be derived from formulas (4.16), (4.17) and (4.18). In case 1 all n picks are located in aisles $\alpha > \alpha_{even}$ and we consider this by the term:

$$\binom{\sum_{i=2x+1}^{\nu} h_i}{n} \tag{4.44}$$

The decision point is also visited if no pick is done in aisle α_{odd} and the last pick to the left of α_{odd} was done in an even aisle and at least one pick is done in an aisle $\alpha > \alpha_{odd}$. We record those cases with:

$$\sum_{y=0}^{n-2}\sum_{z=1}^{n-1-y}\sum_{o=1}^{x}\binom{\sum_{i=1}^{2o-1} h_i}{y}\binom{h_{2o}}{z}\binom{\sum_{i=2x+2}^{\nu} h_i}{n-y-z} \tag{4.45}$$

and require the following conditions for (4.45):

$$y \leq h(2o-1) \qquad z \leq h \qquad (n-y-z) \leq h(\nu-(2x+1))$$

Finally we visit the decision point if at least one pick is done in α_{odd} and at least one pick is done in aisles $\alpha < \alpha_{odd}$ and some picks might be in aisles $\alpha > \alpha_{odd}$. Those cases are included in:

$$\sum_{y=0}^{n-2}\sum_{z=1}^{n-1-y}\binom{\sum_{i=1}^{2x} h_i}{z}\binom{\sum_{i=2x+2}^{\nu} h_i}{y}\binom{h_{2x+1}}{n-y-z} \tag{4.46}$$

The conditions for (4.46) are:

$$z \leq 2xh \qquad (n-y-z) \leq h \qquad y \leq h(\nu-(2x+1))$$

We combine all these cases to get the total number of possibilities to visit decision point C_x:

$$\Gamma_{Cx}^{CB} = \binom{\sum_{i=2x+1}^{\nu} h_i}{n}$$

$$+ \sum_{y=0}^{n-2} \sum_{z=1}^{n-1-y} \sum_{o=1}^{x} \binom{\sum_{i=1}^{2o-1} h_i}{y} \binom{h_{2o}}{z} \binom{\sum_{i=2x+2}^{\nu} h_i}{n-y-z}$$

$$+ \sum_{y=0}^{n-2} \sum_{z=1}^{n-1-y} \binom{\sum_{i=1}^{2x} h_i}{z} \binom{\sum_{i=2x+2}^{\nu} h_i}{y} \binom{h_{2x+1}}{n-y-z} \qquad (4.47)$$

For the special case $x = 0$, there is no aisle left of α_{odd} for $x = 0$:

$$\Gamma_{C0}^{CB} = \binom{\sum_{i=1}^{\nu} h_i}{n} + \sum_{y=0}^{n-2} \sum_{z=1}^{n-1-y} \binom{\sum_{i=2}^{\nu} h_i}{y} \binom{h_1}{n-y-z} \qquad (4.48)$$

In the last step, we consider the number of cases for which C_x is visited but the aisle α_{odd} is not entered. First we count the cases for which all items are located in aisles $\alpha > \alpha_{\overline{even}}$:

$$\binom{\sum_{i=2x+3}^{\nu} h_i}{n} \qquad (4.49)$$

In addition we also consider cases where the decision point is used to skip the adjacent aisles, i.e. the last pick to the left of aisle α_{odd} is done in an even aisle and at least one pick is done in aisles $\alpha > \alpha_{\overline{even}}$. Adapting formula (4.22), we get:

$$\sum_{y=0}^{n-2} \sum_{z=1}^{n-1-y} \sum_{o=1}^{x} \binom{\sum_{i=1}^{2o-1} h_i}{y} \binom{h_{2o}}{z} \binom{\sum_{i=2x+3}^{\nu} h_i}{n-y-z} \qquad (4.50)$$

with conditions:

$$y \leq h(2o-1) \qquad z \leq h \qquad (n-y-z) \leq h(\nu - (2x+2))$$

The overall number of cases $\Gamma_{C_x|\alpha_{next}\neq\alpha_{odd}}$ to be at decision point C_x *and* continuing along the lower cross aisle can be derived according to:

$$\Gamma^{CB}_{C_x|\alpha_{next}\neq\alpha_{odd}} = \binom{\sum_{i=2x+3}^{\nu} h_i}{n}$$

$$+ \sum_{y=0}^{n-2} \sum_{z=1}^{n-1-y} \sum_{o=1}^{x} \binom{\sum_{i=1}^{2o-1} h_i}{y} \binom{h_{2o}}{z} \binom{\sum_{i=2x+3}^{\nu} h_i}{n-y-z} \tag{4.51}$$

$$\Gamma^{CB}_{C_0|\alpha_{next}\neq 1} = \binom{\sum_{i=3}^{\nu} h_i}{n} \tag{4.52}$$

Finally the transition probability at decision point C_x is obtained by:

$$q^{CB}_{C_x,\alpha_{odd}} = 1 - \frac{\Gamma^{CB}_{C_x|\alpha_{next}\neq\alpha_{odd}}}{\Gamma^{CB}_{C_x}} \tag{4.53}$$

$$q^{CB}_{C_0,\alpha=1} = 1 - \frac{\Gamma^{CB}_{C_0|\alpha_{next}\neq 1}}{\Gamma^{CB}_{C_0}} \tag{4.54}$$

4.7 Modeling of Order Picking Times

We will finalize the derivation of the queueing model input parameters by transferring the total order picking time into the input parameters at the respective queueing systems. According to Caron et al. (1998, p. 714), Roodbergen (2001, p. 16), Gudehus (2005, p. 761) or Arnold and Furmans (2009, p. 218) the total order picking time t_{OPT} consists of the following components:

- Travel Time t_{Travel}, which includes the necessary time to walk or drive between picking segments including all cross aisle travel as well as all travel to and from the depot.

- Picking Time t_{pick}, which includes all necessary activities at a picking segment:

- picking an arbitrary quantity of items from two opposite rack columns which can be accessed from the same segment.

- picking items of different handling complexity, e.g. different size and weight.

- all necessary administrative activities at the picking segment, e.g. searching for correct bin, depositing of items onto a handling equipment device, etc.

- Administrative Time t_{Depot}, which includes all necessary activities to deposit a finished order and receive a new one at the depot. This also includes all necessary setup activities.

The total order picking time t_{OPT} can be described by the following formula:

$$t_{OPT} = t_{Depot} + \sum_{i=1}^{n} t_{Pick,i} + t_{Travel,i}$$

Obviously the administrative time t_{Depot} only occurs at the depot, while t_{Pick} only occurs at picking segments within the aisles and t_{Travel} occurs at every segment of the order picking system, i.e. in every queueing system of the queueing network. Before explicitly allocating these time components to the resources of an order picking system (see chapter 4.3) we will consider some general statistical characteristics of times in manual order picking systems.

4.7.1 Statistical Distributions for Times in Manual Order Picking Systems

Most algorithms for the exact solution or approximation of closed queueing networks with blocking assume exponentially distributed service times. With this assumption the system analysis tends to be less complicated. However, in real-life systems generally distributed service times are often regarded as a more realistic assumption (Balsamo, de Nitte Personé and Onvural 2001, p. 5).

Murrell (1962), Dudley (1963), Knott and Sury (1987) and Rall (1998, p.33) state that the distributions of manual work times are typically right-skewed. In particular, Turek and Krengel (2008, p. 5) analyzed operations in manual order picking systems and successfully approximated empirical data by log-normal distributions. They observed coefficients of variation[4] between 0.2 and 0.5 which is in line with the findings of Murrell (1962). Schmidt et al. (2010) focus on manual operations in logistics systems and report similar results.

We assume that the picking time t_{Pick} has a logarithmic normal distribution (log-normal distribution). The distribution is closely related to the well known normal distribution, as a random variable X has a log-normal distribution if its logarithm $\ln(X)$ has a normal distribution. Random variable $X \sim \mathcal{LN}(\mu, \sigma^2)$ has the following probability density function:

$$f(x) = \begin{cases} \frac{1}{\sqrt{2\pi}\cdot\sigma} \cdot \frac{1}{x} \cdot e^{-\frac{(\ln x - \mu)^2}{2\sigma^2}} & \text{if } x > 0 \\ 0 & \text{if } x \leq 0 \end{cases}$$

The expected value and variance are defined as:

$$E(X) = e^{\mu + \frac{1}{2}\sigma^2} \tag{4.55}$$

$$Var(X) = (e^{\sigma^2} - 1) \cdot e^{2\mu + \sigma^2} \tag{4.56}$$

Vice versa, μ and σ^2 are defined as:

$$\mu = \ln(E(X)) - \frac{1}{2}\ln\left(1 + \frac{Var(X)}{E(X)^2}\right) \tag{4.57}$$

$$\sigma^2 = \ln\left(1 + \frac{Var(X)}{E(X)^2}\right) \tag{4.58}$$

Figure 4.11 shows the probability density function of X with $E(X) = 10$ and $Var(X) = 50$, i.e. variability $c^2 = 0.5$.

The shape of the density function reflects some important observations we can make in manual order picking systems:

---

[4]The coefficient of variation is a measure of spread that normalizes the standard deviation over the mean service time. It is widely used in existing queueing theoretic solution and approximation algorithms.

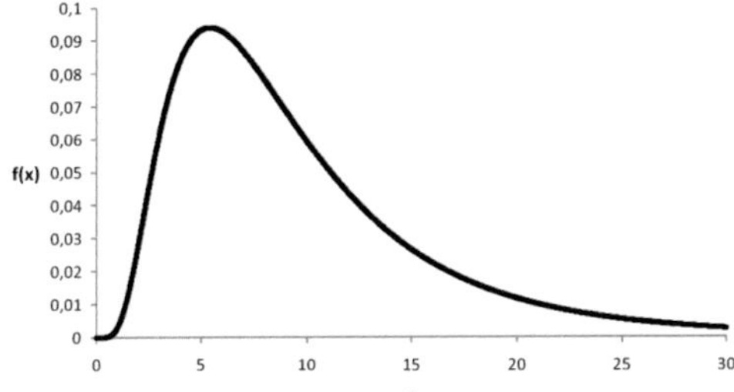

Figure 4.11: Density of a log-normal distribution

- No negative working times will occur and the density assumes positive values for positive realizations only.

- For each activity in the picking system we will have a minimum working time. Thus the mode of the distribution will not come for values close to zero but for values closer to the mean value.

- Considering the mode of the distribution of working times, there is a higher potential for time values to deviate upwards than to deviate downwards. In real-life systems, times larger than the mode are feasible and might include extra work, e.g. opening of packages or administrating a missing parts situation. Times lower than the mode happen rarely. This observation is mirrored in the right skewness of the log-normal distribution.

- Depending on the scope of different activities, the variability of the working time can have different values and does not necessarily have to be equal to 1.

In the following we will assume that both the administrative time t_{Depot} and the picking time t_{Pick} follow a log-normal distribution. Consequently in our queueing model each system representing the depot or a segment within the aisle will have generally distributed service times t_s. We need to characterize each queueing system i with a service rate μ_i and a variability c_i^2. According to Arnold and Furmans (2009, p. 21) we can use the mean value $E(t_s)$ and variance $Var(t_s)$ to calculate the input parameters for queueing system i as:

$$\mu_i = \frac{1}{E(t_{s,i})} \qquad c_i^2 = \frac{Var(t_{s,i})}{E(t_{s,i})^2} \tag{4.59}$$

4.7.2 Service Time at the Depot

In chapter 4.3 we defined that the depot was represented by one segment which in turn was modeled as one elementary queueing system. The activities to pass through this system thus includes both travel *and* the administrative activities. While the latter will be generally distributed, we will assume constant travel times within the whole system. Therefore the travel time at the depot t_{Travel} will be a deterministic variable T with:

$$E(T) = \frac{Length\ of\ one\ segment}{Picker\ Velocity} \qquad Var(T) = 0$$

The administrative time t_{Depot} will be modeled by a random variable D with $D \sim \mathcal{LN}(\mu_D, \sigma_D^2)$. Note that the expected value $E(D)$ and variance $Var(D)$ can be derived using formulas (4.55) and (4.56). The overall service time at the depot $t_{s,Depot}$ will be the sum of travel time and administrative time:

$$t_{s,Depot} = t_{Travel} + t_{Depot}$$

The expected value is given by:

$$E(t_{s,Depot}) = E(T) + E(D)$$

Because the variables are independent we get:

$$Var(t_{s,Depot}) = Var(T) + Var(D) = Var(D)$$

Ultimately we obtain the service time input parameters for the depot by using formulas (4.59):

$$\mu_{Depot} = \frac{1}{E(t_{s,Depot})} = \frac{1}{E(T) + E(D)} \tag{4.60}$$

$$c_{Depot}^2 = \frac{Var(t_{s,Depot})}{E(t_{s,Depot})^2} = \frac{Var(D)}{[E(T) + E(D)]^2} \tag{4.61}$$

4.7.3 Service Time at a Cross Aisle

Each segment of the cross aisle will be modeled by an elementary queueing system. At a cross aisle pickers will never pick but just travel. We define the travel time t_{Travel} as a deterministic variable T with:

$$E(T) = \frac{Length\ of\ one\ segment}{Picker\ Velocity} \qquad Var(T) = 0$$

Thus the service time at a cross aisle $t_{s,CA}$ equals the travel time t_{Travel} and the expected value will be:

$$E(t_{s,CA}) = E(T)$$

The input parameters for the queueing model are then easily derived:

$$\mu_{CA} = \frac{1}{E(t_{s,CA})} = \frac{1}{E(T)} \tag{4.62}$$

$$c_{CA}^2 = \frac{Var(t_{s,CA})}{E(t_{s,CA})^2} = \frac{Var(T)}{E(T)^2} = \frac{0}{E(T)^2} = 0 \tag{4.63}$$

4.7.4 Service Time within an Aisle

The service time at each within-aisle queueing system has to include both travel time *and* picking time. As before, the travel time t_{Travel} will be defined by a deterministic variable T. The duration of the picking process is influenced by two factors: the stopping probability $p_{stop,i}$, indicating the probability that a picker has to stop and pick items at segment i and the stochastic picking time $t_{pick,i}$. The service time within an aisle generally is:

$$t_{s,WA} = t_{Travel} + p_{stop,i} \cdot t_{pick,i} \tag{4.64}$$

We will derive the service time $t_{s,WA}$ based on a sample set of orders. Let $x_{i,j}$ be a binary variable that indicates whether a pick has to be done at segment i for order number j. An arbitrary set of M orders is given by:

		Segments							
		1	2	3	4	...	$\nu h - 2$	$\nu h - 1$	νh
Orders	1	0	1	0	1	...	0	0	1
	2	0	0	0	1	...	1	0	1
	3	1	1	0	0	...	0	0	0
	4	0	1	0	1	...	1	1	0
	...	...	...	...	...	...	...	...	...
	$M - 2$	1	0	1	0	...	0	1	0
	$M - 1$	0	0	1	0	...	0	1	1
	M	0	0	0	0	...	1	0	0

As each order has exactly n picks, we can define the following constraint:

$$\sum_{i=1}^{\nu h} x_{i,j} = n \qquad \forall j = 1, 2, ..., M$$

If we assume a *random storage policy*, we can calculate the probability that a pick occurs at segment i, according to:

$$p_{pick,i} = \frac{\sum_{j=1}^{M} x_{i,j}}{M} = \frac{n}{\nu h} \qquad \forall i = 1, 2, ..., \nu h \qquad (4.65)$$

If a *class-based storage policy* with classes $(A,B,C,...,L^{CB})$ is in use, the probability that a pick occurs at segment i in an arbitrary order is given by:

$$p_{pick,i} = \frac{\sum_{j=1}^{M} x_{i,j}}{M} = \frac{\gamma_{p,picks} \cdot n}{\gamma_{p,seg} \cdot \nu h} \qquad \forall i = 1, 2, ..., \nu h \qquad (4.66)$$

For both storage location assignment policies, we have n picks per order, thus:

$$\sum_{i=1}^{\nu h} p_{pick,i} = n$$

Formulas (4.65) and (4.66) equal the probability to actually stop at segment i *only* if *all* segments of the system are visited in one order picking tour. Thus, to determine the stopping probability $p_{stop,i}$, we also have to include the visit ratios e_i into our considerations, because they include the information on how often a segment is visited in relation to the number of visits at the depot. The stopping probability will be calculated according to:

$$p_{stop,i} = p_{pick,i} \cdot \frac{1}{e_i}$$

For better understanding, imagine an order picking system with 10 aisles and traversal routing policy with aisle skipping. Let us assume that aisle 10 holds items of the C-class and let the above formulas result in very low probabilities $p_{pick,i}$ for the segments in aisle 10. If an order picker actually visits aisle 10, he has pick(s) in aisle 9 and/or 10. Otherwise he would never had entered aisle 10 in the first place. Consequently in this case the probability to stop and pick at a segment in aisle 10 is much higher than simply the probability that a segment of aisle 10 is on an arbitrary order.

For *random storage policy*, we can therefore define the expected value of the stopping probability as:

$$E(p_{stop,i}) = \frac{n}{\nu h} \cdot \frac{1}{e_i} \qquad \forall i = 1, 2, ..., \nu h \qquad (4.67)$$

For *class-based storage policy*, we get:

$$E(p_{stop,i}) = \frac{\gamma_{p,picks} \cdot n}{\gamma_{p,seg} \cdot \nu h} \cdot \frac{1}{e_i} \qquad \forall i = 1, 2, ..., \nu h \qquad (4.68)$$

As $x_{i,j}$ is a binary variable we can obtain the variance $Var(p_{stop,i})$ by using the second moment:

$$E(p_{stop,i}^2) = 1^2 \cdot p_{stop,i} + 0^2 \cdot (1 - p_{stop,i}) = p_{stop,i}$$

The variance is given by:

$$Var(p_{stop,i}) = E(p_{stop,i}^2) - E(p_{stop,i})^2$$

Thus for *random storage* we have:

$$Var(p_{stop,i}) = \frac{n}{\nu h} \cdot \frac{1}{e_i} - \left(\frac{n}{\nu h} \cdot \frac{1}{e_i}\right)^2 \qquad (4.69)$$

For *class-based storage* we have:

$$Var(p_{stop,i}) = \frac{\gamma_{p,picks} \cdot n}{\gamma_{p,seg} \cdot \nu h} \cdot \frac{1}{e_i} - \left(\frac{\gamma_{p,picks} \cdot n}{\gamma_{p,seg} \cdot \nu h} \cdot \frac{1}{e_i} \right)^2 \tag{4.70}$$

The stochastic picking time $t_{pick,i}$ is given by a random variable P with $P \sim \mathcal{LN}(\mu_P, \sigma_P^2)$. We assume that $E(P)$ and $Var(P)$ are known from an empirical analysis using a suitable methodology, e.g. REFA or MTM (Schlick et al. 2010, p. 664).

In order to obtain the service time $t_{s,WA}$ as defined in formula (4.64) we need to handle a product of two random variables, namely $p_{stop,i} \cdot t_{pick,i}$. We can assume that these two variables are independent, because the length of an arbitrary picking activity itself does not depend on the decision to pick or not. For a specific visit to segment i the picking time will be 0 if $x_{i,j} = 0$ because then $x_{i,j} \cdot t_{pick,i} = 0$.

The expected value of the product of two independent random variables is easily calculated by (Fahrmeir et al. 2003, p. 245):

$$E(X \cdot Y) = E(X) \cdot E(Y)$$

and thus:

$$E(p_{stop,i} \cdot P) = E(p_{stop,i}) \cdot E(P) \tag{4.71}$$

The variance of a product of two independent random variables is given by (Goodman 1960):

$$Var(X \cdot Y) = Var(X) \cdot Var(Y) + E(X)^2 \cdot Var(Y) + E(Y)^2 \cdot Var(X)$$

and thus:

$$Var(p_{stop,i} \cdot P) = Var(p_{stop,i}) \cdot Var(P) \tag{4.72}$$
$$+ E(p_{stop,i})^2 \cdot Var(P) + E(P)^2 \cdot Var(p_{stop,i})$$

To fully characterize the overall service time, the travel time t_{Travel} is added and we get:

$$E(t_{s,WA}) = E(T) + E(p_{stop,i}) \cdot E(P)$$

The travel time is independent from any picking activity and therefore:

$$Var(t_{s,WA}) = Var(T) + Var(p_{stop,i} \cdot P)$$

The input parameters for the queueing model are now fully defined as:

$$\mu_{WA} = \frac{1}{E(t_{s,WA})} = \frac{1}{E(T) + E(p_{stop,i}) \cdot E(P)} \tag{4.73}$$

$$c^2_{WA} = \frac{Var(t_{s,WA})}{E(t_{s,WA})^2} = \frac{Var(T) + Var(p_{stop,i} \cdot P)}{[E(T) + E(p_{stop,i}) \cdot E(P)]^2} \tag{4.74}$$

For different segments i within an aisle we can assume different characteristics values μ_P and σ_P^2. Thus different segments can potentially have different service times and the model enables to consider different item quantities as well as various sizes and weights.

4.8 Validation of Traversal Routing Policy and Service Time Calculation

For the purpose of validating our formulas, we applied them to a range of different order picking systems. The values calculated by the formulas were compared to the results of simulation. We combined several specifications of aisles (6,8,10,12,14,16), warehouse shape factors[5] ($\frac{1}{3},\frac{2}{3},1$), pick densities[6] (5%,10%,20%) and storage location assignment policies (random storage, medium-skewed class-based-storage and high-skewed class-based storage) for a total of 162 experiments.

All experiments were simulated with 50 replications, each over a period of 50 days. Both factors play a key role in obtaining valid results from a simulation model. According to Hillier (1997, p. 773), it is impossible

---

[5]Following existing literature (e.g. (Petersen 1997, p. 1103) or (Hwang, Oh and Lee 2004, p. 3883)), the warehouse shape factor is defined as the ratio between length L and width W of the order picking system.

[6]A pick density of 5% states that on each tour an order picker has to stop at 5 locations in a system with 100 locations.

to assess a generally valid number of replications. The simulation time is of paramount importance as each system requires a certain time until it has reached a steady state. We should further note that in contrast to aggregated values such as throughput, the routing and service time parameters appear not to be as robust with respect to the simulation time[7] (Rall 1998, p. 142). We chose 50 for both repetitions and simulation days and found that a further increase will only marginally improve accuracy. Hence 50 is a good compromise between accuracy and computational effort.

We briefly review our analysis steps for the parameter *service times*. For each of the 162 experiments, we first calculate the mean error *for each individual* node $i = 1, 2, ..., N$ according to:

$$\varepsilon_{E(t_{s,WA,i})} = \frac{E(t_{s,WA,i}) - t_{s,WA,Simulation,i}}{t_{s,WA,Simulation,i}}$$

We then calculate the mean error *over all* N nodes in the respective experiment by:

$$\bar{\varepsilon}_{E(t_{s,WA})} = \frac{\sum_{i=1}^{N} \varepsilon_{E(t_{s,WA,i})}}{N}$$

Solely using this formula might result in negative and positive errors of the individual nodes leveling each other, falsely suggesting an accurate result. Therefore, we also calculate a 99%-confidence interval for the mean error using[8]:

$$\text{Lower Bound: } \bar{\varepsilon}_{E(t_{s,WA})} - 2.5758 \cdot \sqrt{\frac{\frac{1}{N-1} \cdot \sum_{i=1}^{N} \left(\varepsilon_{E(t_{s,WA,i})} - \bar{\varepsilon}_{E(t_{s,WA})}\right)^2}{N}}$$

$$\text{Upper Bound: } \bar{\varepsilon}_{E(t_{s,WA})} + 2.5758 \cdot \sqrt{\frac{\frac{1}{N-1} \cdot \sum_{i=1}^{N} \left(\varepsilon_{E(t_{s,WA,i})} - \bar{\varepsilon}_{E(t_{s,WA})}\right)^2}{N}}$$

---

[7] In our models this is especially true whenever we have segments that are very rarely visited (e.g. class C). Consequently it takes relatively long until a sufficient number of events have taken place at this segment, ensuring a steady state value of the parameter.

[8] As all experiments have at least $N > 30$ nodes, an approximative confidence interval is used (Fahrmeir et al. 2007, p. 389)

For each experiment, the confidence interval characterizes the range of deviations over all nodes. We repeated these steps for the other two parameters $c^2_{WA,i}$ and e_i respectively. Table 4.1 summarizes the minimum and maximum values that resulted from all 162 experiments:

	Mean	Confidence Interval	
$E(t_{s,WA})$	$\bar{\varepsilon}_{E(t_{s,WA})}$	Lower Bound 99%	Upper Bound 99%
Minimum	-0.39%	-0.98%	-0.22%
Maximum	0.35%	0.11%	0.85%
c^2_{WA}	$\bar{\varepsilon}_{c^2_{WA}}$	Lower Bound 99%	Upper Bound 99%
Minimum	-0.52%	-0.95%	-0.32%
Maximum	0.63%	0.45%	1.34%
e_i	$\bar{\varepsilon}_e$	Lower Bound 99%	Upper Bound 99%
Minimum	-0.60%	-0.90%	-0.35%
Maximum	0.57%	0.34%	0.98%

Table 4.1: Validation of visit ratios and characteristic service time values

The table reads as follows: over all 162 experiments the maximum mean error for the parameter c^2_{WA} was 0.63%. Note that the values of one line do not correspond to each other as they most likely stem from different experiments. For example, 0.45% was the largest lower bound that we could identify for all of the 162 intervals that were calculated. Likewise, 1.34% was the largest upper bound we observed for the parameter c^2_{WA} when considering all 162 experiments. We can see that the mean errors are always smaller than 1% and the maximum upper bound of all confidence intervals is 1.34%, thus the simulation values differ only marginally from our calculated values.

4.9 Chapter Conclusion

The input parameters of the queueing model are now fully defined. We should note that this modeling approach is applicable to a wide range of manual order picking systems. It does not matter whether pickers walk with a picking cart or drive on a forklift truck. Only the size of the segments und thus the travel time t_{Travel} would change. Secondly, both low-level and high-level systems can be modeled by altering the picking time t_{pick}.

Table 4.2 summarizes the transformation by showing how the parameters of the order picking system and the queueing model are related to each other.

Order Picking System		Queueing Model	
Layout	ν, h	N	Number of nodes
		m	Single-server queueing systems
Narrow Aisles	W_{Aisles}	B_i	No buffers, BAS protocol
Routing Policy	One-way traversal		
Number of picks	n	e_i	Visit ratios
Storage Policy	Random, Class-Based		
Times	t_{Pick}, t_{Depot}	μ, c^2	Service rate, variability
Pickers	K	K	Customers

Table 4.2: Parameter transformation - summary

By applying the presented model, continuous time queueing theory can basically be used to analyze the effects of congestion in order picking systems. In contrast to existing analytical approaches (see chapter 3.2), the model provides an extended scope because assumptions are less restrictive.

Nevertheless, we still need a solution technique to actually calculate performance indicators such as throughput. The assumptions on the queueing model obviously have a great influence on the applicability of existing solution techniques as some seemingly minor details often have significant impact on the procedure of these techniques. The presented transformation results in the following characteristics:

- Closed queueing network with a fixed number of customers
- Single-Server elementary queueing systems without any buffers
- Blocking-After-Service protocol
- General service times

In the next chapter, we will present existing methods of queueing theory, discuss the problems when applying them to the above modeling approach and develop a new approximation method.

5 Calculation of Performance Indicators for Closed Zero-Buffer Queueing Networks

The consideration of blocking phenomena complicates calculation procedures for closed queueing networks and many standard approaches are not applicable anymore. In order to calculate throughput for the presented queueing model of manual order picking systems, we will first analyze in chapter 5.1 to what extent existing approaches can be used. Subsequently, we will focus on the accuracy of a promising approach - Rall's method - in chapter 5.2. This will give us some ideas on how to design our own approximation approach, which will be presented in chapter 5.3.

5.1 Existing Algorithms

We will give a brief overview of remaining approaches and will discuss those methods influencing the development of our own approach in more detail.

5.1.1 Overview

If we concentrate on existing approaches for closed queueing networks, we only find a limited number of procedures. By additionally considering Blocking-After-Service protocol and general service times, only four potential approaches remain. To the best of our knowledge, no other algorithms exist which are applicable to networks underlying the assumptions mentioned before.

An approach developed by Bouhchouch et al. (1996) uses decomposition techniques to study closed tandem queueing networks. It therefore tends to be computationally expensive with regards to large networks. More importantly though the approach is only suitable for cyclic configurations, i.e. each system has only one predecessor and one successor. Hence visit ratios $e_i \neq 1$ cannot be considered.

Vroblefski et al. (2000, 2005) also use a decomposition approach. They model networks based on the flow of kanbans through a production line. In (2000) cyclic network configurations can be analyzed. An extension to telecommunication networks with switches is presented in (2005). However, the network structures analyzed are specifically chosen as to represent telecommunication networks (e.g. ring, star or lattice topologies) and they include buffers behind a server for pooling finished data packets and synchronization stations to realise the forwarding. The approach is therefore not suitable for models of order picking systems.

A major contribution to the analysis of queueing networks with blocking was presented by Akyildiz. Some basics to calculate throughput for closed queueing networks with blocking and exponential service times were presented in the *matching state space algorithm* in (1987) and (1988b). For a given network with blocking these approaches attempt to find an equivalent non-blocking network with the same state space cardinality. Ultimately this will lead to a reduction in the number of customers from K to K^*. The throughput for a blocking network with K customers can then be approximated by calculating the throughput of a non-blocking network with K^* customers. Note that these algorithms are exact for networks with two nodes and approximative for more than two nodes.

Akyildiz (1988a) extended the matching state space algorithm to networks with general service times (*Akyildiz's integrated approach*). The author presents a procedure to find an equivalent non-blocking network that best approximates the number of active servers in the blocking network. Again this is done by reducing the number of customers to K^*. Along with the original structure of the blocking network this number is then used as an input in *Marie's method* for closed queueing networks with general service times and unlimited buffer sizes (Marie 1980). Some relative large errors were reported in the original literature (Akyildiz 1988a, p. 301) which might be a reason that the algorithm has gained relatively small attention from the scientific community. Furthermore rather small networks with up to five nodes were analyzed. Onvural (1990, p. 107) states that the original matching state space algorithm works well for networks with exponentially distributed service times. However, accuracy is worse when the procedure is combined with Marie's method for networks with general service times.

Rall's integrated approach (1998) is based on the idea formulated in (Akyildiz 1988a). Rall extended this idea by introducing a correction factor f_{Rall}, which modifies K^*, the result of Akyildiz's matching state space algorithm thus adapting it to non-exponential networks. The modified number of customers is then used as an input to Marie's method. The original algorithms of Akyildiz (1987) (1988b) (1989) are applicable to closed queueing networks with finite buffers and exponentially distributed service times. The proposed convolution operation will in most cases lead to a wrong reduction of the number of customers if service times are general. For $c^2 < 1$ there will be less blocking situations in the network compared to $c^2 = 1$ (Rall 1998, p. 127). This is because the flow of customers through the network is less obstructed. For the extreme case of $c^2 = 0$ blocking situations might only occur at merging points - otherwise customers will flow through the network in a certain tact. Furthermore we can assume that for $c^2 > 1$ there will be more blocking situations than for $c^2 = 1$ because the network will behave more dynamically. Rall (1998, p. 132) reported that the procedure suggested by Akyildiz (1988b) in fact produces a correct throughput trend but still underestimates ($c^2 < 1$) or overestimates ($c^2 > 1$) the real throughput. From this Rall assumed that the value of K^* has to be modified adequately before using it as an input value for Marie's method. Multiplying K^* with the correction factor f_{Rall} will result in K^{**}, the input for Marie's method:

$$K^{**} = f_{Rall} \cdot K^* \tag{5.1}$$

Due to the underlying assumptions, both Akyildiz's matching state space algorithm (1988b) as well as Marie's method (1980) are not directly applicable to our models of manual order picking systems. Still, we will provide a short description of these algorithms in chapters 5.1.2 and 5.1.4 as they are used in both Akyildiz's integrated approach and Rall's integrated approach respectively. We will also discuss the derivation of Rall's correction factor in chapter 5.1.3. The understanding of these three methods is important for chapter 5.2, where we will discuss the accuracy of Rall's integrated approach when applied to order picking systems. We will show how the particular methods of this framework influence the throughput approximation error, giving us some valuable ideas for our own approach.

5.1.2 Akyildiz's Matching State Space Algorithm

This algorithm presented by Akyildiz in (1987 (for two nodes), 1988b (for more than two nodes) and 1989 (for multiple servers)) can be applied to closed queueing networks with exponential service times. Consider two closed queueing networks which are similar except for the size of the buffers. One network has finite buffers (blocking network), the other network has infinite buffers (non-blocking network). The basic idea of Akyildiz is that the throughput of these two networks is approximately equal if the state spaces of both networks have the same cardinality. For the non-blocking network we can calculate the number of states Z for N nodes and K customers by applying the binomial coefficient:

$$Z = \binom{N + K - 1}{N - 1}$$

The state space of the blocking network has to consider both possible apportionments of customers as well as blocking states. Akyildiz suggests a procedure to estimate the size of the state space by an efficient convolution operation.

Deriving the State Space of Blocking Networks

The state space vector Z^* is calculated by the convolution of N capacity vectors:

$$Z^* = Z_1 \otimes Z_2 \otimes \ldots \otimes Z_N$$

If vector C is the result of the convolution of vectors A and D, the kth element is given by:

$$C(k) = \sum_{j=0}^{k} A(j) \cdot D(k-j)$$

For each node in the network the $(K+1)$-dimensional capacity vector Z_i is given by:

$$Z_i = \begin{bmatrix} a_i(0) \\ a_i(1) \\ \vdots \\ a_i(K) \end{bmatrix}$$

where $a_i(k)$ are the values of a binary function:

$$a_i(k) = \begin{cases} 1 & \text{for} \quad k = 0, 1, ..., (B_i + \sum_{j=1 \wedge q_{ji}>0}^{N} m_j) \\ 0 & \text{otherwise} \end{cases} \tag{5.2}$$

The value of 1 represents either a regular or a blocking state and B_i is the capacity of system i and m_j is the number of servers at stations with a positive transition probability q_{ji}. Once the state space vector Z^* is derived, we can obtain the size of the state space for K customers by reading the $(K+1)^{st}$ element of Z^*. By definition the first element of Z^* equals 0.

Transformation into a Non-Blocking Network

Following Akyildiz's basic idea we have to find a non-blocking network such that

$$Z \approx Z^*(K+1)$$

The binomial coefficient specifies that Z only depends on N and K. Since N must not change, we have to alter K to K^* to obtain a Z approximately equal to $Z^*(K+1)$:

$$Z^*(K+1) \approx \binom{N + K^* - 1}{N - 1} \tag{5.3}$$

K^* is always smaller than or equal to K because the state space of a blocking network is always smaller than or equal to the state space of a non-blocking network.

According to Akyildiz, we can observe the following:

$$\lambda_{Blocking}(K) \approx \lambda_{Non-Blocking}(K^*)$$

To calculate the throughput $\lambda_{Non-Blocking}(K^*)$ any algorithm for product-form networks can be used (e.g. Mean Value Analysis (Reiser and Lavenberg 1980) or the convolution algorithm (Buzen 1973)).

Summary and Critique

We briefly summarize the steps of Akyildiz's matching state space algorithm:

1. Calculate the resulting vector Z^* by convolving N capacity vectors Z_i. The size of the state space of the blocking network is given by $Z^*(K + 1)$

2. The size of the state space for a non-blocking network, Z, is given by the binomial coefficient. Use N and K^* in the binomial coefficient and alter K^* such that $Z \approx Z^*(K + 1)$.

3. Use K^* in any algorithm for product-form networks to approximate the throughput $\lambda(K)$ of the blocking network.

As mentioned, the method is exact for two nodes. For more than two nodes, the original literature reports stable numerical results. Akyildiz (1988b) analyzed 150 queueing networks. Most of these were small networks with four or less stations. The largest network evaluated had 8 stations. Balsamo et al. (2001, p. 154f.) report that the relative error of this method may easily reach 25% even for small networks. Similar observations were made by Dallery and Frein (1989).

5.1.3 Rall's Correction Factor and Integrated Approach

Influencing Parameters

Rall (1998) suggested a correction factor depending on both the customer density of the network ω, and a representative network variability c_R^2. He assumed that blocking will be worse with increasing variability of the service processes and increasing customer density ω of the network. The latter is defined as:

$$\omega = \frac{K}{\sum_{i=1}^{N} B_i}$$

The representative network variability is calculated using the nodes' individual variabilities which are weighted with the following characteristic values:

- Visit ratio e_i: the influence of variability c_i^2 will increase with increasing e_i

- Node utilization ρ_i: a node with high utilization ρ_i will likely cause more blocking situations than a node with small utilization

- Number of servers m_i: parallel servers can reduce the mean queue length $\overline{k}$ and thus calm the network

- Ratio of mean queue length and system capacity $\frac{\overline{k_i}}{B_i}$: a high ratio will increase the probability of blocking situations thus influencing the network variability

The weighting factors ρ_i and $\overline{k_i}$ have to be calculated by applying any kind of product form network to the underlying network. For homogeneous networks, c_R^2 equals c^2 of any node. For heterogeneous networks, c_R^2 is calculated according to:

$$c_R^2 = \frac{\sum_{i=1}^{N} e_i \cdot \rho_i^2 \cdot 1/\sqrt{m_i} \cdot \overline{k_i}/B_i \cdot c_i^2}{\sum_{i=1}^{N} e_i \cdot \rho_i^2 \cdot 1/\sqrt{m_i} \cdot \overline{k_i}/B_i} \tag{5.4}$$

Rall analyzed a five-node network with $\sum_{i=1}^{N} B_i = 16$ and 12 different variabilities in the range of 0.25 to 5. Based on the empirical observations of the resulting 240 configurations, Rall (1998, p. 129) derived the following formula for the correction factor:

$$f_{Rall} = 1 - \omega + \omega \cdot \frac{1}{\sqrt[3]{c_R^2}} \tag{5.5}$$

It is assumed that f_{Rall} grows linearly with the customer density ω and over-proportionally with the representative network variability. Note that f_{Rall} can be either larger or smaller than 1. For $c_R^2 < 1$ the correction factor is larger than 1 and the customer reduction made by Akyildiz will be partly reversed. For $c_R^2 > 1$ the correction factor is smaller than 1 and the customer reduction made by Akyildiz will be intensified further.

Summary

The steps of Rall's framework can be summarized as follows:

1. Calculate the utilization ρ_i and the mean number of customers in the queue $\overline{k_i}$ for each node in the network using any product-form algorithm, e.g. mean value analysis

2. Calculate the representative network variability c_R^2 with formula (5.4)

3. Determine a reduced number of customers K^* using the matching state space algorithm by Akyildiz

4. Calculate the correction factor f_{Rall} with formula (5.5)

5. Calculate the updated number of customers K^{**} using formula (5.1)

6. Use K^{**} as an input in Marie's method to calculate the throughput of the blocking network with general service times. As step 5 might result in a non-integer value for K^{**}, Marie's algorithm will be run twice using $\lfloor K^{**} \rfloor$ and $\lceil K^{**} \rceil$ respectively. The final result $\lambda(K)$ is obtained by weighting the two throughputs:

$$\lambda_{Rall}(K^{**}) = a \cdot \lambda_{Marie}(\lfloor K^{**} \rfloor) + b \cdot \lambda_{Marie}(\lceil K^{**} \rceil)$$

where the parameters a and b are defined as follows:

$$a = \lceil K^{**} \rceil - K^{**} \qquad b = K^{**} - \lfloor K^{**} \rfloor$$

5.1.4 Marie's Method

The method of Marie calculates throughput for a closed queueing network with infinite buffers and general service times. The approach was introduced by Marie in (1979) and (1980). Useful descriptions are also given in Bolch et al. (1998, p. 452f.) and (Rall 1998, p. 102f.). Marie and Pellaumail (1983) give some more detail on the underlying assumptions and Marie et al. (1982) discuss the computational aspects of the method.

The author suggests an iterative scheme successively switching between a macro view and a micro view. In the macro view, the whole network is analyzed using load-dependent service rates $\mu_i(k)$. The result of this macro view, the load-dependent arrival rates $\lambda_i(k)$, are then used in the micro view to calculate the conditional throughput rates $\nu_i(k)$ and the state probabilities $\pi_i(k)$. The conditional throughput rates $\nu_i(k)$ are used as load-dependent service rates $\mu_i(k)$ in the next iteration loop. The algorithm terminates once two stopping conditions are fulfilled. The variables are connected by the following condition:

$$\nu_i(k) \cdot \pi_i(k) = \lambda_i(k-1) \cdot \pi_i(k-1) \tag{5.6}$$

It states that the probability of a customer leaving a node in state k is equal to the probability that a customers arrives at the same node in state $(k-1)$. Based on this, a continuous time Markov Chain with birth rate $\lambda_i(k)$ and death rate $\nu_i(k)$ can be assigned to node i. The state probabilities can be calculated as follows:

$$\pi_i(k) = \pi_i(0) \prod_{j=0}^{k-1} \frac{\lambda_i(j)}{\nu_i(j+1)} \qquad \forall \qquad k = 1, ..., K \tag{5.7}$$

$$\pi_i(0) = \frac{1}{1 + \sum_{j=1}^{K} \prod_{k=0}^{j-1} \frac{\lambda_i(k)}{\nu_i(k+1)}} \tag{5.8}$$

For the derivation of $\nu_i(k)$, elementary $\lambda_{(}k)|G|1|K - FCFS$ nodes are analyzed. Therefore the method of Marie seems well suited to calculate characteristic values for closed queueing networks with general service times. Other algorithms for this type of network, namely the *response time preservation approach (RTP)* of Agrawal et al. (1984) or the *generalized summation method* as described in Bolch et al. (1998, p. 463), do not consider load-dependent arrival rates. This assumption can lead to inaccurate results in material flow systems, especially for smaller number of customers (Furmans 2000, p. 104). The method of Marie therefore represents the closest approximation to networks consisting of $G|G|1$ nodes, which we will need in the later analysis.

Macro View - Calculating Load-Dependent Arrival Rates

In order to obtain load-dependent arrival rates, the given non-product form network with generally distributed service times is transformed into a product-form network with exponentially distributed service times. For each node a substitute network (SN) is created by short-circuiting node i. This is done by adopting the visit ratios: for node i we set $e_i = 1$; for all other nodes $j = 1, ..., N; j \neq i$ we divide e_j by the original e_i. The load-dependent arrival rates $\lambda_i(k)$ can be derived from the throughput at the short-circuit of the substitute network.

If k customers reside in node i then exactly $(K-k)$ customers are located at the other nodes of the substitute network. The throughput of the substitute network with $(K-k)$ customers therefore is the arrival rate at node i with k customers (Bolch et al. 1998, p. 452):

$$\lambda_i(k) = \lambda_{SN}(K - k), \qquad \text{for k} = 0,1,...,\text{(K-1)}$$

By definition $\lambda_i(K) = 0$ as there will not be any arrivals at node i if all customers reside at node i. To obtain $\lambda_{SN}(K - k)$ any algorithm for a product-form network can be used. Note that these algorithms need load-dependent service rates as an input. These stem from the conditional throughput rates $\nu_i(k)$ which are calculated in the micro view.

Micro View - Calculating Conditional Throughput Rates

The computation of $\nu_i(k)$ depends on the variability of the service time at node i, namely c_i^2. The following cases have to be considered:

- $c_i^2 = 1$: The conditional throughput rates $\nu_i(k)$ are equal to the initial service rate μ_i
- $c_i^2 \geq 0.5$: A Cox-2 model is chosen to approximate the general service times
- $c_i^2 = \frac{1}{m} \pm \varepsilon$ with $m = \{m \mid m \in \mathbb{N} \wedge m \geq 3\}$: An Erlang model with m exponential phases is used
- All other cases: A Cox-m model is chosen

A description of these procedures is given in (Bolch et al. 1998, p. 454). The derivation of service times for order picking systems as described in chapter 4.7 will most likely result in variabilities $c_i^2 \geq 0.5$. Therefore the Cox-2 model will be of special interest.

Once the $\nu_i(k)$ are derived, the state probabilities $\pi_i(k)$ are calculated by using formulas (5.7) and (5.8). The throughput at any node i is then given by:

$$\lambda_i = \sum_{k=1}^{K} \pi_i(k) \cdot \nu_i(k) \tag{5.9}$$

Stopping Conditions and Node Throughput

After each micro view, the following two stopping conditions are checked. First, we assess whether the sum of the mean number of customers over all nodes equals the total number of customers K:

$$\left| \frac{K - \sum_{i=1}^{N} \sum_{k=1}^{K} k \cdot \pi_i(k)}{K} \right| \leq \varepsilon \tag{5.10}$$

Secondly it is checked whether the throughput rates computed with formula (5.9) are consistent with the topology of the network. Therefore we compare the throughput rates to the normalized average throughput rate with:

$$\left| \frac{\frac{\lambda_j}{e_j} - \frac{1}{N} \sum_{i=1}^{N} \frac{\lambda_i}{e_i}}{\frac{1}{N} \sum_{i=1}^{N} \frac{\lambda_i}{e_i}} \right| < \varepsilon \qquad \forall j = 1, ..., N \tag{5.11}$$

Once both conditions are fulfilled, the algorithm terminates and the node throughput rates are obtained from equation 5.9.

Summary and Critique

Let us briefly summarize the steps of Marie's method:

1. Create substitute networks by short-circuiting node i for $i = 1, ..., N$. Initialize the load-dependent service rates $\mu_i(k)$ by the service rates of the original network.

2. Macro view: calculate load-dependent arrival rates $\lambda_i(k)$ by analyzing the substitute networks by means of an algorithm for product-form networks.

3. Micro view: calculate the conditional throughput rates $\nu_i(k)$ and the state probabilities $\pi_i(k)$ by making use of the $\lambda_i(k)$ obtained in the macro view. The approximation scheme depends on the variability of node i.

4. Assess the stopping conditions (5.10) and (5.11). If not fulfilled set the conditional throughput rates as the new load-dependent service rates of the substitute network, i.e. $\mu_i(k) = \nu_i(k)$ and return to the macro view (step 2).

Bolch et al. (1998, p. 462) report that the method delivers quite accurate results. The convergence of the method has not been proven. However, we experienced no cases of diverging behavior when using reasonable values of ε for the stopping conditions. Both Bolch et al. (1998, p. 458) and Rall (1998, p. 111) suggest values for ε to be in a range of 10^{-3} to 10^{-4}. We should note that we could not identify any literature in which the method was applied to large queueing networks. Sadre et al. (2007, p. 167) note that the method appears to be suitable for rather small models with multiple-servers. Souza et al. (1986, p. 421) argue that no error bound has been shown. A discussion of potential error sources can be found in Dallery and Cao (1992, p. 58). The authors identify the step of approximating the conditional throughput rates $\nu_i(k)$ as a potential error source, because it is based on a state-dependent Poisson process, i.e. assuming that arrivals occur independently.

5.2 Accuracy of Rall's Integrated Approach

Concerning the accuracy of his approach, Rall (1998, p. 134) reports stable results for cyclic networks. However, for networks with arbitrary configurations, accuracy decreases and throughput tends to be overestimated (Rall 1998, p. 135). The derivation of the correction factor f_{Rall} as well as the validation of the framework were conducted based on rather small queueing networks. With regard to the analysis of larger order picking systems we therefore must run a separate validation cycle in order to assess the framework's accuracy.

We will use the original framework for a set of order picking systems and compare the analytical results to simulation. The discussion of the approximation quality will be followed by some considerations on how it is influenced by the frameworks' three main steps (Akyildiz's matching state space algorithm, calculation of the correction factor, method of Marie).

5.2.1 Experiment Design

To get a general idea on the approximation quality, we applied Rall's framework to a large set of order picking systems, which will be called *experiment set 1* in the following. We first analyzed the parameters which were used in the well-known publications presented in chapter 3 to get a first idea for our selection. With regard to practical applications and computational efficiency (as we have to use simulation) we selected the following range of parameters:

- Number of aisles ν: 6, 8, 10, 12, 14, 16
- Warehouse shape factor WSF: $\frac{1}{3}$, $\frac{2}{3}$, 1 (the shape factor is defined as the ratio $\frac{L}{W}$)
- Pick Density: $\frac{1}{20}$, $\frac{1}{10}$, $\frac{1}{5}$ (the pick density is defined as the ratio $\frac{n}{\nu \cdot h}$)
- Storage Location Assignment Policy: Random, medium-skewed class-based, high-skewed class-based. We assume both class-based policies have three classes with the A-class occupying 20% of the segments, the B-class occupying 30% and the C-class occupying

50% of the segments. For medium-skewed, the picks will be distributed on these classes by 50/30/20%, i.e. 50% of the picks will be located in 20% of the segments. For high-skewed, this distribution will be 80/15/5%.

- Picker Density (based on segments) ω_{seg}: $\frac{1}{30}$, $\frac{2}{30}$, $\frac{3}{30}$, ... , $\frac{10}{30}$
- Service time $t_{s,Depot}$: 1, 5, 15 (seconds)
- Variability c^2_{Depot}: 0.5
- Picking time t_{Pick}: 15 (seconds)
- Variability Picking time c^2_{Pick}: 0.5

Experiment set 1 covers a wide spectrum of scenarios which can be found in real-life manual order picking systems. Based on these parameters the size of *experiment set 1* is 4860 experiments.

For the accuracy analysis we use a simulation model built with Plant Simulation software. For validation purposes, we also used two other programs to independently implement the order picking systems: our own JAVA code as well as the JAVA Modelling Tools[1] (for selected experiments).

5.2.2 Overall Approximation Quality

Accuracy for All Experiments

We assume that simulation results are sufficiently close to the exact values. Thus, the relative error ε of the analytical approach corresponds to the relative deviation between analytical results and simulation:

$$\varepsilon = 100 \cdot \frac{\lambda_{\text{analytic}} - \lambda_{\text{simulative}}}{\lambda_{\text{simulative}}}$$

Figure 5.1 shows the histogram of deviations between Rall's framework and simulation. The probability mass function of the deviation is right skewed, i.e. the framework of Rall has a tendency to overestimate throughput. More than half of the experiments in *experiment set 1* had

---

[1] Open source software developed by Politecnico di Milano

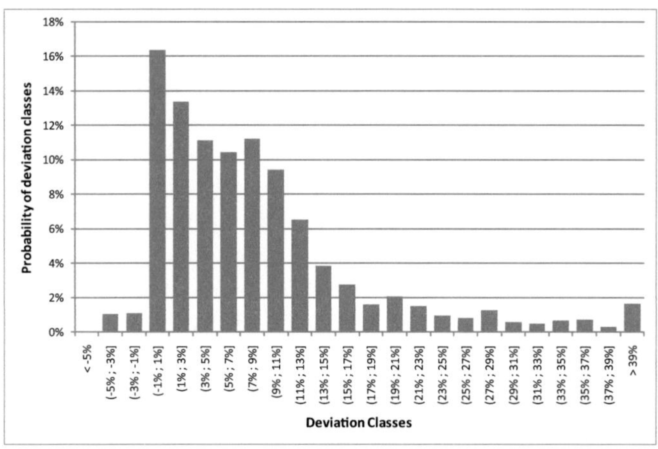

Figure 5.1: Deviation between Rall's framework and simulation

a relative error larger than $|5\%|$ and overall we can record the following observations:

- $P(\varepsilon \leq |5\%|) = 42\%$
- $P(\varepsilon \leq |10\%|) = 68\%$
- $P(\varepsilon \geq 20\%) = 12\%$

Accuracy Based on Certain Parameters

We want to identify the key parameters that mainly cause the inaccuracy of the existing approach. Unsurprisingly due to the heterogeneous structure of the queueing networks, trends are not of linear nature. We calculated Pearson's coefficient of correlation between all of our parameters and the overall deviation and the largest value was 0.08 for picker density ω_{seg}.

As we further analyze the influence of different parameters, we separated the experiments by parameter classes. A parameter class is made up by all experiments having a certain parameter value, e.g. all experiments with picker density $\omega_{seg} = 0.1$ make up one class. We use the ANOVA procedure of Kruskal and Wallis (1952) to test whether the experiments from different parameter classes have different medians[2]. We combine these results with a simple comparison of mean deviations for the different classes. We demonstrate our procedure for the picker density classes, i.e. 10 classes because of 10 different picker density values.

We first perform a Kruskal-Wallis-ANOVA to find out that deviations from different classes do not come from the same population at a significance level of 95%, i.e. the errors for picker density $\omega_{seg} = 0.1$ are different to those of $\omega_{seg} = 0.2$. Comparing the mean error values within the classes we get:

$$\bar{\varepsilon}_{\omega_{seg}=0.033} = 0.9\% \quad \dots \quad \bar{\varepsilon}_{\omega_{seg}=0.1} = 4.1\%\dots$$

$$\bar{\varepsilon}_{\omega_{seg}=0.2} = 9.4\% \quad \dots \quad \bar{\varepsilon}_{\omega_{seg}=0.333} = 19.6\%$$

We repeated this procedure for all parameters to come up with the following patterns:

- deviation between the Rall's integrated approach and simulation will be higher for higher levels of picker density ω_{seg}, higher skewness of the class-based distribution as well as for larger networks (larger number of aisles and/or larger warehouse shape factor)

- deviation between Rall's integrated approach and simulation does not seem to be determined by the representative network variability c_R^2, the pick density as well as the service time $t_{s,Depot}$

5.2.3 Influence of Akyildiz's Matching State Space Algorithm

We will analyze two aspects to evaluate the influence of this algorithm. For selected order picking systems we will first compare the results of the approach to simulation. We will also discuss the impact of zero-buffer queueing systems on the convolution operation.

---

[2]We could not use standard ANOVA procedures as deviations were not normally distributed within most of the parameter classes.

We altered the original *experiment set 1* such that all $c^2 = 1$, thus having exponentially distributed service times at all nodes. Figure 5.2 shows the deviation between Akyildiz's approach as described in section 5.1.2 and simulation structured by different storage location assignment policies.

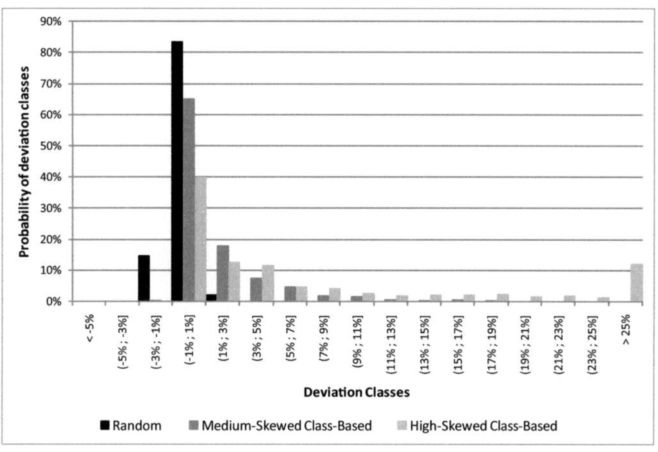

Figure 5.2: Deviation between Akyildiz's algorithm and simulation

For random storage, the algorithm works very well as all experiments have an error $\varepsilon \leq |5\%|$. For medium-skewed class-based storage though, we can identify some experiments producing errors larger than $|5\%|$. The results for high-skewed class-based storage are unsatisfactory as almost 40% of experiments have an error $\varepsilon \geq 5\%$ and we also have a non-negligible number of experiments with errors $\varepsilon \geq 25\%$. It seems as if the method is not capable of correctly modeling level-2, level-3, etc. blocking situations which occur more frequently in class-based storage.

One reason for this might be the inaccuracy of the method for arbitrary, i.e. non-cyclic network configurations, which we undoubtedly have in order picking systems. By applying formula (5.2) we increase the number of entries in the capacity vectors Z_i for each preceding node. Compared to cyclic networks, this will lead to an increased size of the state space with every node merging different streams. Consequently the number of customers K^* will also increase when using formula (5.3) (Rall 1998, p. 153).

Another reason might be the procedure of the convolution operation itself. Imagine an order picking system with N nodes and a traversal routing policy without aisle skipping. We thus have a cyclic configuration and all nodes have one successor and the following capacity vectors:

$$Z_i = \begin{pmatrix} 1 \\ 1 \\ 1 \\ 0 \\ 0 \\ \vdots \end{pmatrix} = \begin{pmatrix} 1 \\ 1 \\ 1 \end{pmatrix}$$

We begin the convolution operation with:

$$\begin{pmatrix} 1 \\ 1 \\ 1 \end{pmatrix} \otimes \begin{pmatrix} 1 \\ 1 \\ 1 \end{pmatrix} = \begin{pmatrix} 1 \\ 2 \\ 3 \\ 2 \\ 1 \end{pmatrix}$$

Continuing with the third node, we get:

$$\begin{pmatrix} 1 \\ 2 \\ 3 \\ 2 \\ 1 \end{pmatrix} \otimes \begin{pmatrix} 1 \\ 1 \\ 1 \end{pmatrix} = \begin{pmatrix} 1 \\ 3 \\ 6 \\ 7 \\ 6 \\ 3 \\ 1 \end{pmatrix}$$

We identify the following pattern: after k convolutions, the maximum entry of the resulting vector is always given for element $(k + 2)$. In a network with N nodes, we apply the convolution operation $(N - 1)$ times. Thus the maximum entry of the resulting vector will be given for element $(N-1+2) = (N+1)$ which represents the state space cardinality for N customers. Note that in a network with N zero-buffer nodes, the maximum number of customers K_{max} will be N. This translates into the following observation: for increasing number of customers K, the trends of state space cardinality, number of customers K^* and throughput will all reach a degree of saturation for N customers. In particular, as N is the maximum number of customers, this means that these curves will never decrease in the range of $K = 1, 2, ..., K_{max}$.

The number of merging elements is limited in order picking systems and therefore, even though we will have capacity vectors Z_i with four "1" elements, the above observations remain more or less unchanged. If we assume high-skewed class-based storage and focus on throughput, we will easily reach a degree of saturation for number of customers $K \ll N$ and the reduction of customers $K \to K^*$ will be too small. Figure 5.3 highlights this observation. Almost 80% of experiments had no reduction at all and the maximum reduction was 3.57%.

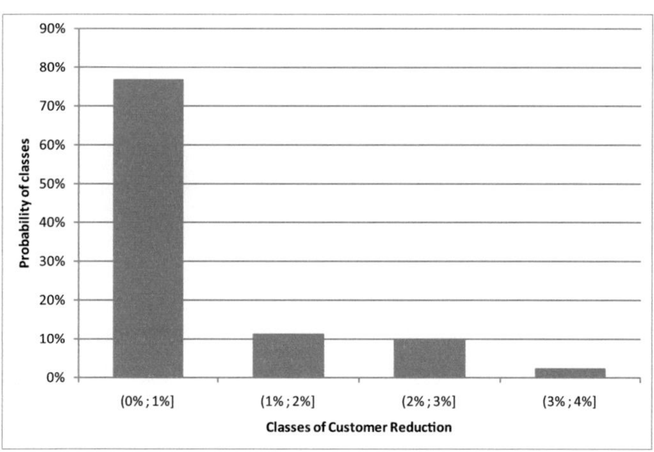

Figure 5.3: Reduction of customers calculated by Akyildiz's algorithm

This analysis has revealed that the algorithm of Akyildiz is not applicable to the queueing models of order picking systems that we have presented. Especially the convolution operation in zero-buffer networks can lead to false results as these systems are behaving differently than those with limited buffers[3].

5.2.4 Influence of the Correction Factor

In Rall's original framework the correction factor f_{Rall} is embedded between the algorithm of Akyildiz and the method of Marie. Both methods are approximative, i.e. possibly causing considerable errors. Therefore we have to isolate these two error sources in order to assess the accuracy of the correction factor f_{Rall}.

Any approximation error stemming from Marie's method could be erased by applying simulation in the last step of Rall's framework. However, the error of Akyildiz would still be incorporated. Let us briefly review the original intention of f_{Rall}. It updates the result of Akyildiz's algorithm depending on the network's variability c_R^2, i.e. in case of $c_R^2 < 1$ the reduction in the number of customers is partly reversed, for $c_R^2 > 1$, the reduction in the number of customers is intensified.

For a particular closed queueing network, we first assume that $c_i^2 = 1 \ \forall \ i$. We can then calculate the true throughput by simulation and find a K_{true}^* for which

$$\lambda_{Blocking,Sim}^{c_i^2=1}(K) \approx \lambda_{Non-Blocking,MVA}(K_{true}^*)$$

K_{true}^* then is the reduced number of customers that should have been the correct result of Akyildiz's algorithm. We can also find a K_{true}^{**} for which

$$\lambda_{Blocking,Sim}^{c_i^2\neq1}(K) \approx \lambda_{Non-Blocking,Sim}(K_{true}^{**})$$

K_{true}^{**} then is the reduced number of customers that should have been the correct result after applying f_{Rall}. Let K_{Rall}^{**} be the actual number calculated by

$$K_{Rall}^{**} = K_{true}^* \cdot f_{Rall}$$

---

[3]This statement is also supported by the results of a personal discussion with the author (Akyildiz 2009).

Comparing K^{**}_{Rall} to K^{**}_{true} will give us an idea, whether the correction factor f_{Rall} would have worked had we eliminated all other sources for errors. The error $\varepsilon_{K^{**}}$ is calculated by

$$\varepsilon_{K^{**}} = \frac{K^{**}_{Rall} - K^{**}_{true}}{K^{**}_{true}}$$

and the errors for different parameter scenarios are shown in table 5.1.

SLAP	$P(\varepsilon_{K^{**}} \leq |5\%|)$	$P(\varepsilon_{K^{**}} \leq |10\%|)$
Overall	83%	94%
Random	96%	98%
Medium-Skewed Class-Based	86%	96%
High-Skewed Class-Based	65%	88%

Table 5.1: Error in calculating K^{**} using Rall's correction factor

Overall we can record a fairly good accuracy. If we categorize the results by storage location assignment policy, we can easily see that f_{Rall} performs good for random storage and performs worse with increasing skewness of the class-based storage.

Overall for random storage a linear trend of the correction factor seems plausible. For class-based storage however, we have to consider the following thoughts. In class-based storage blocking situations might quickly emerge even for very small picker densities ω_{seg}. This is because pickers tend to be located in a cycle within the whole network, e.g. the first two aisles. Let $\mathcal{J}$ be a cycle with j nodes ($j = 2, 3, ..., j_{max} | j_{max} < N$). A cycle can hold at most $\sum_{\forall j \in \mathcal{J}} B_j$ customers and we can define cycle picker densities $\omega_{\mathcal{J}}$ as:

$$\omega_{\mathcal{J}} = \frac{K}{\sum_{\forall j \in \mathcal{J}} B_j} \tag{5.12}$$

By solely considering the original picker density ω_{seg} instead of $\omega_{\mathcal{J}}$ some blocking situations cannot be considered and the throughput might be overestimated. These cycles are not specifically defined by f_{Rall}.

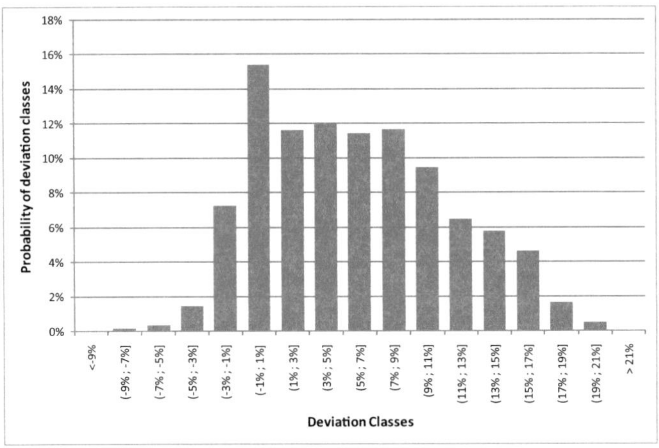

Figure 5.4: Deviation between Marie's method and simulation

5.2.5 Influence of Marie's Method

As the approach of Marie is an approximation method, we have to quantify the accuracy to make a statement on how far it influences the overall accuracy of Rall's integrated approach. For that reason we altered *experiment set 1* such that all nodes have unlimited buffers. We then first compared the results of Marie to those of simulation. In a second step we analyzed how big the deviations correspond to certain input parameters.

Figure 5.4 shows the deviation between the method of Marie and simulation for experiment set 1. Throughput is mostly overestimated, apart from 16% of experiments resulting in an underestimation. The error is existent, however it is fairly moderate with almost 75% of experiments having an error $\varepsilon \leq |10\%|$ and the maximum error being $\varepsilon = 20.61\%$.

In line with the procedure for the overall accuracy, we classified the experiments by input parameters to identify those parameters with a strong interrelation to the output values. We found the representative network variability c_R^2 to have a strong influence on deviation. Again we ran the

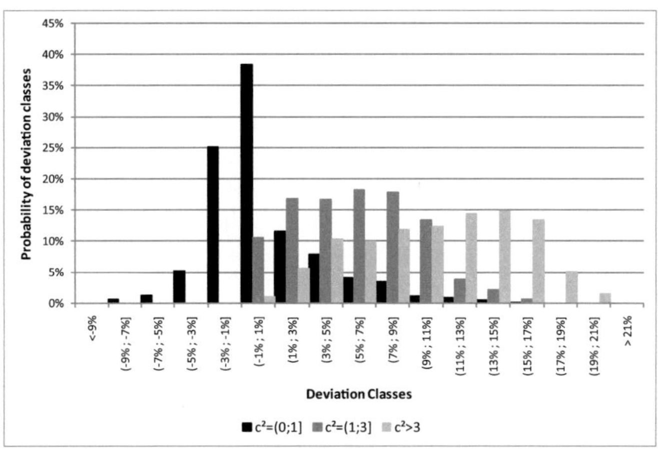

Figure 5.5: Deviation between Marie's method and simulation classed by c_R^2

Kruskal-Wallis-ANOVA procedure to confirm that deviations are different for different classes of c_R^2. In particular we see that mean deviation is increasing for increasing c_R^2:

$$\bar{\varepsilon}_{c_R^2 \in (0;0.75]} = -0.1\% \qquad \bar{\varepsilon}_{c_R^2 \in (1.25;1.75]} = 4.9\% \qquad \bar{\varepsilon}_{c_R^2 \in (4.75;5.5]} = 12.9\%$$

Furthermore figure 5.5 illustrates the context as the distribution of deviation is given for different levels of c_R^2.

Other input parameters influencing the overestimation are picker density ω_{seg} as well as the number of nodes N in the network. Again the mean deviation is increasing for increasing input parameters, namely:

$$\bar{\varepsilon}_{\omega_{seg}=0.033} = 1.7\% \quad \ldots \quad \bar{\varepsilon}_{\omega_{seg}=0.1} = 4.7\%$$
$$\bar{\varepsilon}_{\omega_{seg}=0.2} = 6.6\% \quad \ldots \quad \bar{\varepsilon}_{\omega_{seg}=0.333} = 7.4\%$$

$$\bar{\varepsilon}_{N=57} = 0.1\% \quad \ldots \quad \bar{\varepsilon}_{N=145} = 7.2\%$$
$$\bar{\varepsilon}_{N=222} = 8.2\% \quad \ldots \quad \bar{\varepsilon}_{N=424} = 9.3\%$$

The root cause for these deviations can be traced back to the calculation of conditional throughput rates $\nu_i(k)$. They describe the rate at which a certain state is left after a service has been finished. As the states are characterized by the number of customers in the system, $\nu_i(k)$ effectively describes the throughput of a queueing system depending on the number of customers in that system. It is needed to derive steady-state probabilities as in formula (5.7). For Cox-2 distributions, the conditional throughput rates $\nu_i(k)$ depend solely on the parameters of the Cox-2 distribution and the state-dependent arrival rates. We show in the appendix A that the parameters proposed by Marie (1980) produce correct approximations of desired values μ and c^2. Thus the state-dependent arrival rates $\lambda_i(k)$ remain as a source for error. They are based on a Poisson process, i.e. interarrival times are assumed to be exponentially distributed. However, in closed queueing networks this assumption is violated if service times are non-exponential.

We used simulation to measure the true steady-state values of $\nu_i(k)$, $\lambda_i(k)$ and $\pi_i(k)$ and were able to confirm Marie's steady-state condition (see formula 5.6). Inserting the values in formula (5.9) resulted in the same throughput which we calculated with simulation. When using the method of Marie, the steady-state condition will also hold and the method will converge after some iteration steps. However, in each step the false $\lambda_i(k)$ will lead to false values of $\nu_i(k)$ and the combination of both will lead to false $\pi_i(k)$ (see formula 5.7) and thus false throughput. The iteration scheme "is aware" of this and tries to consider this error by using the $\nu_i(k)$ as new load-dependent service rates $\mu_i(k)$ in the next iteration step. Nevertheless the throughput will not necessarily adjust to the true throughput.

The false estimate of $\lambda_i(k)$ is mainly based on variability in the network. $\lambda_i(k)$ will differ more from the true values, with more nodes having non-exponentially distributed service times. To illustrate this we selected one experiment[4] to show a characteristic trend in figure 5.6. For clarification we assume all nodes have the same variability c^2, thus making the network slightly more homogeneous than an actual order picking system. Marie's method will underestimate throughput for $c^2 < 1$ because the $\lambda_i(k)$ will

---

[4] $\nu = 10$, $h = 10$, $n = 10$, $\omega_{seg} = 20\%$, $t_{s,Depot} = 5s$, SLAP=Random

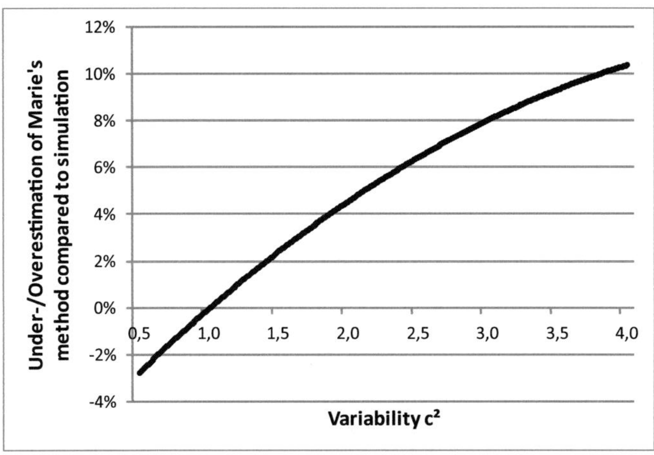

Figure 5.6: Influence of variability on accuracy of Marie's method

be too small as the throughput of a product-form algorithm (assuming $c^2 = 1$) will be too small. Marie's method will overestimate throughput for $c^2 > 1$ because the $\lambda_i(k)$ will be too big as the throughput of a product-form algorithm (assuming $c^2 = 1$) will be too big. In order picking systems, variabilities will be different for each node and we have to use the aggregated c_R^2 to describe them. Nevertheless the effect is also visible, especially if c_R^2 is strongly influenced by those nodes mainly affecting throughput, e.g. nodes with high visit ratios.

Overall, we have found that results of the approximation are mostly within an error range of $|10\%|$. The error will increase with increasing c_R^2, ω_{seg} and skewness of the class-based storage.

Finally, we should note that the accuracy of Rall's framework cannot be improved by replacing Marie's method with the response time preservation approach (RTP) of Agrawal et al. (1984) or the generalized summation method presented by Bolch et al. (1998, p. 463). For small picker densities $\omega_{seg} \leq 10\%$, Marie's method performed better. This is because the use of $\lambda(k)|G|1|K - FCFS$-nodes by Marie allows to consider state-dependent arrival rates, thus providing a more realistic approximation of arrival processes within the closed network. Note that this result is in line with Furmans (2000, p. 104). For other picker densities, all three algorithms calculate more or less the same results. Both RTP and the generalized summation method are computationally less expensive than Marie's method. Still, because of its advantages for small ω_{seg}, Marie's method appears to be the best choice when analyzing order picking systems.

5.2.6 Implications for a New Approach

Both the approach of Akyildiz (1988a) as well as Rall (1998) attempt to estimate throughput losses based on blocking situations by calculating the throughput in a network with unlimited buffers after having adjusted the number of customers in the network. Especially the integrated approach of Rall already produces partly accurate results when applied to models of order picking systems. To further improve the accuracy, we will develop a refined procedure which is based on the following learnings:

- In zero-buffer networks, as we use them, the matching state space algorithm of Akyildiz will not reduce the number of customers in the majority of cases. This is due to the definition of the convolution operation. We will therefore not consider this method as a part of our new approximation approach.

- Rall's correction factor seems to work well for some parameter scenarios, e.g. random storage. However, its structure does not seem appropriate if pickers are distributed in a way that certain aisles of the system have very high picker densities while other aisles have very low picker densities. We introduce a new correction factor f_{ZB} which is built on some other functional relation. It incorporates the effects of zero-buffers compared to networks with unlimited buffers.

- The method of Marie will remain a part of our new approximation method even though it produces some approximation errors by estimating conditional throughput rates on the basis of a Poisson arrival process. We will introduce a distinct correction factor f_{Marie} to handle these errors.

5.3 A New Integrated Approach

5.3.1 Basic Idea

With the aid of one specific example order picking system[5] we will illustrate the basic idea of our new integrated approach, which is based on the observations in the previous chapter.

Zero-Buffer vs. Unlimited Buffer Networks

For the chosen queueing network, we will first analyze the throughput trends when comparing the simulation of a zero-buffer network to the simulation of a network with unlimited buffers for *random storage policy* (see figure 5.7).

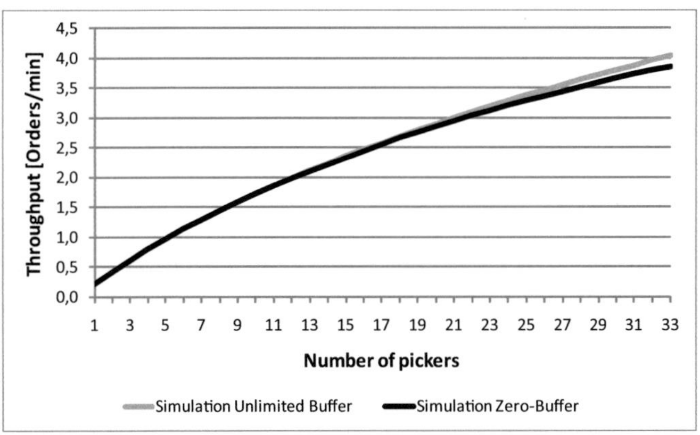

Figure 5.7: Throughput trends for random storage policy: zero-buffer vs. unlimited buffer

---

[5]The following parameters were used: $\nu = 10$, $h = 10$, $n = 10$, $K = (1, 2, ..., 33)$, $t_{s,Depot} = 1s$, $c^2_{Depot} = 0.5$, $t_{s,WA} = 15s$ and $c^2_{WA} = 0.5$

We can easily see that there is little or no difference at all between the two trends so we can assess that for random storage the introduction of limited buffers has little or no impact on throughput.

It is important to realize that this does *not* mean that there is no throughput loss due to congestion. It simply means that congestion losses are almost fully incorporated in the model with unlimited buffers.

For an explanation, we will emphasize that for level-1-blocking situations, the size of the buffer does *not* matter. As long as no third picker is involved, it is irrelevant if the blocked picker waits in the buffer or in the server. The waiting time caused by the blocking situation thus is a *normal* waiting time as already considered in networks with unlimited buffer size.

If a third picker is involved and a level-2-blocking situation evolves, there can be more than just *normal* waiting time. We can imagine a situation in which the third picker wants to use the server which is blocked by the second picker. This server would have been available, had the second picker already moved to the buffer of the succeeding system. In the zero-buffer case, this delay is reducing throughput. However, it is only of importance if the third picker would have actually used the second server for picking. If he just needs to walk then the delay is considerably small and will have only marginal influence on overall throughput.

The number of blocking situations of higher order in general hugely depends on picker density, i.e. we will have more level-2-3-...-blocking situations if more pickers work in the system. Thus for higher ω we should see more throughput losses that actually stem from the non-existing buffers and the resulting level-2-3-...-blocking situations. Still though these losses seem rather small as for random storage the pickers distribute more or less evenly over the whole system. If pickers meet, the probability is high that no more than two are involved.

We also analyzed the same network for *class-based storage* and expectedly obtained a different result. Figure 5.8 shows the respective throughput trends when comparing the simulation of a zero-buffer network to the simulation of a network with unlimited buffers.

For a limited number of pickers in the system ($K \leq 10$), the throughput of both zero-buffer and unlimited buffer networks is approximately equal

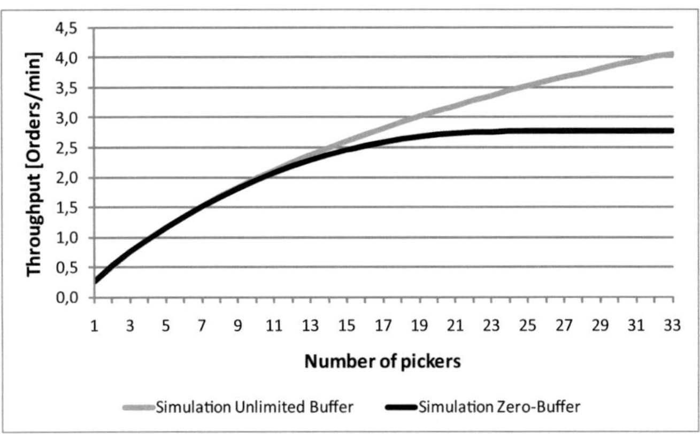

Figure 5.8: Throughput trends for class-based storage policy: zero-buffer vs. unlimited buffer

($\epsilon \leq 1\%$). Like for random storage, congestion almost entirely stems from level-1-blocking. Beyond that point, the two trends begin to drift apart and for $K = 33$ throughput of the zero-buffer network is 32% smaller than throughput in the unlimited buffer network. Comparing the zero-buffer networks for random storage (figure 5.7) and class-based storage (figure 5.8), we also notice that for $K < 18$ throughput is higher for class-based storage and for $K \geq 18$ throughput is higher for random storage. Hence, there is some critical level for the number of pickers K where a change of storage policy is advisable. This is because for class-based storage, level-2-3-...-blocking situations occur for smaller ω as in random storage. The explanation for this is quite straightforward. In class-based storage, pickers will not evenly distribute over the whole system but are likely to be located in an aisle with many picks, i.e. an aisle located closer to the depot. The picker density in these aisles ω_α will be higher than in aisles located far from the depot. Consequently there will be more level-2-3-...-blocking situations.

We will now present the basic idea of the correction factor f_{ZB}. As an example, let us calculate the throughput of the given system for $K = 20$

pickers, i.e. $\lambda_{OPS}(K = 20)$. The throughput of the order picking system is measured in *picked order lines per period of time* and thus:

$$\lambda_{OPS}(K) \approx \lambda_{ZB}(K) \cdot n$$

We search for a number of pickers K^+ such that the throughput in a network with unlimited buffers and K^+ pickers approximately equals the throughput in a zero-buffer network with K pickers:

$$\lambda_{UB}(K^+) \approx \lambda_{ZB}(K)$$

Thus the number of customers is reduced from K to K^+ as shown in figure 5.9.

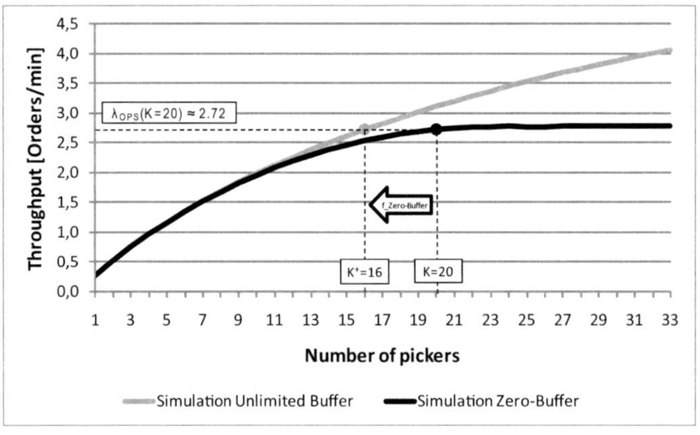

Figure 5.9: Effect of correction factor f_{ZB} for class-based storage

The graph shows the trends for class-based storage. In such a configuration, we have to reduce the number of pickers from $K = 20$ to approximately $K^+ = 16$. We put the emphasis on the term *approximately* because K^+ does not necessarily have to be an integer value (see chapter 5.3.2). Using $K^+ = 16$ in a network with unlimited buffers results in a good throughput approximation of throughput of the zero-buffer network:

$$\lambda_{UB}^{Class-Based}(16) = 2.717 \approx 2.716 = \lambda_{ZB}^{Class-Based}(20)$$

The correction factor f_{ZB} is generally defined as follows:

$$f_{ZB} = \frac{K^+}{K} \qquad (5.13)$$

In the example, f_{ZB} would therefore be $\frac{16}{20} = 0.8$. Note that for random storage (see figure 5.7) the correction factor f_{ZB} would be approximately 1, because no reduction of pickers is necessary to get a useful approximation of throughput:

$$\lambda_{UB}^{Random}(20) = 2.89 \approx 2.85 = \lambda_{ZB}^{Random}(20)$$

Unlimited Buffer Networks vs. Method of Marie

By multiplying the original number of customers K with the correction factor f_{ZB}, we get K^+ which can be used in a network with unlimited buffers. Chapter 5.2.5 has shown that the well-known analytical methods, e.g. the method of Marie, might over- or underestimate throughput of such networks.

Following the idea of f_{ZB}, we will define a correction factor f_{Marie} that will adjust the number of pickers from K^+ to K^{++} in a way that:

$$\lambda_{Marie}(K^{++}) \approx \lambda_{UB}(K^+)$$

Consider figure 5.10 for an example. For class-based storage the number of customers is further reduced from $K^+ = 16$ to $K^{++} = 14$ to have approximately equal throughputs in a simulation of the network with unlimited buffers and the analytical calculation using the method of Marie:

$$\lambda_{Marie}^{Class-Based}(14) = 2.66 \approx 2.716 = \lambda_{UB}^{Class-Based}(16)$$

The correction factor f_{Marie} is generally defined as follows:

$$f_{Marie} = \frac{K^{++}}{K^+} \qquad (5.14)$$

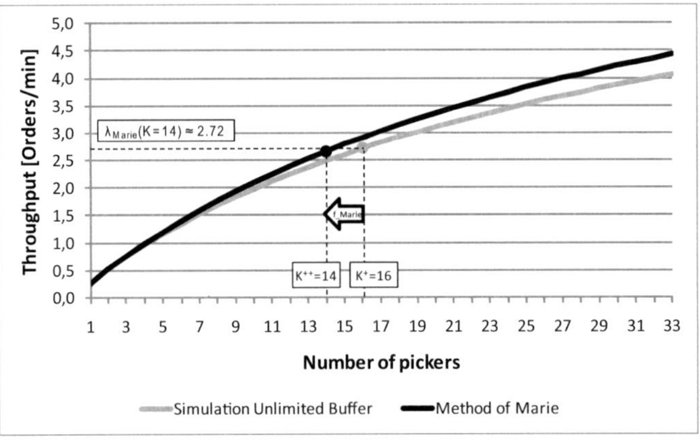

Figure 5.10: Effect of correction factor f_{Marie} for class-based storage

In the example, f_{Marie} would therefore be $\frac{14}{16} = 0.875$. Note that for random storage, we would need to reduce the number of pickers from $K^+ = 20$ to $K^{++} = 17$, i.e. $f_{Marie} = 0.85$, to get a useful approximation:

$$\lambda_{Marie}^{Random}(17) = 2.84 \approx 2.89 = \lambda_{UB}^{Random}(20)$$

Combination of f_{ZB} and f_{Marie}

We will briefly rerun the example including all steps for the case of class-based storage. The throughput of the order picking system (with $n = 10$) is obtained by simulating a zero-buffer network:

$$\lambda_{OPS}(20) \approx \lambda_{ZB}(20) \cdot 10 = 27.2$$

We obtain K^+ by:

$$K^+ = K \cdot f_{ZB} = 20 \cdot 0.8 = 16$$

Applying f_{Marie} results in K^{++}:

$$K^{++} = K^+ \cdot f_{Marie} = 16 \cdot 0.875 = 14$$

We finally use $K^{++} = 14$ in the method of Marie to get:

$$\lambda_{Marie}(14) = 2.66$$

With this we obtain a useful approximation of throughput in the order picking system:

$$\lambda_{OPS}(20) \approx \lambda_{ZB}(20) \cdot 10 = 27.2 \approx 26.6 = \lambda_{Marie}(14) \cdot 10$$

Note that this approximation is a 2.2% underestimation of real throughput. This is based on the fact that only integer values were used in this example. We will show in the following chapter that both K^+ and K^{++} can attain non-integer values, ultimately resulting in better approximations. Note that for $K < 3$, f_{ZB} is not considered as there will be at most level-1-blocking situations which are covered by Marie's method.

5.3.2 Procedure for the Derivation of True Correction Factors

The derivation of both f_{Marie} and f_{ZB} follows the same procedure. We will present in detail the approach for f_{Marie}. The factor f_{ZB} can be derived accordingly. Again, the example of the previous chapter is used as we illustrate the derivation of the true value f_{Marie}^{true} for a single experiment.

We assume that f_{ZB} has already reduced $K = 20$ to $K^+ = 16$. For all $K = (1, 2, 3, ..., 33)$ we obtain $\lambda_{UB}(K)$ from simulation and $\lambda_{Marie}(K)$. From these 33 pairs of values we now need to find the throughput values that are closest to $\lambda_{UB}(16) = 2.717$ and find the corresponding values of K that will result in a throughput close to $\lambda_{UB}(16)$. Table 5.2 shows some of the closest values.

From the table we see that the throughput value $\lambda_{UB}(16) = 2.717$ is best approximated by either $K = 14$ or $K = 15$ because:

$$(\lambda_{Marie}(14) = 2.663) \leq (\lambda_{UB}(16) = 2.717) \leq (\lambda_{Marie}(15) = 2.792)$$

Instead of choosing one integer value of K to be our K^{++}, we use a linear combination of the two closest K to get a non-integer value K_{true}^{++} which would result in the closest approximation of $\lambda_{UB}(K^+)$.

K	$\lambda_{UB}(K)$	$\lambda_{Marie}(K)$
1	0.271	0.270
...	...	...
12	2.245	2.390
13	2.373	2.530
14	2.489	2.663
15	2.605	2.792
16	2.717	2.914
17	2.820	3.032
18	2.916	3.145
...	...	...
33	4.059	4.424

Table 5.2: Derivation of correction factors - search for closest throughput values

Let K_x be the value of K leading to the closest approximation $min[\lambda_{UB}(K^+) - \lambda_{Marie}(K_x)]$ and let K_{x+1} denote the value of K leading to the closest approximation $min[\lambda_{Marie}(K_{x+1}) - \lambda_{UB}(K^+)]$. Then the linear combination can be calculated according to:

$$K_{true}^{++} = \frac{\lambda_{UB}(K^+) - K_{x+1} \cdot \lambda_{Marie}(K_x) + K_x \cdot \lambda_{Marie}(K_{x+1})}{\lambda_{Marie}(K_{x+1}) - \lambda_{Marie}(K_x)} \quad (5.15)$$

Based on this we can calculate the true value of the correction factor f_{Marie}^{true}:

$$f_{Marie}^{true} = \frac{K_{true}^{++}}{K^+} \quad (5.16)$$

Using the numbers of the example we get:

$$K_{true}^{++} = \frac{\lambda_{UB}(16) - 15 \cdot \lambda_{Marie}(14) + 14 \cdot \lambda_{Marie}(15)}{\lambda_{Marie}(15) - \lambda_{Marie}(14)}$$

$$= \frac{2.717 - 15 \cdot 2.663 + 14 \cdot 2.792}{2.792 - 2.663}$$

$$= 14.41$$

and the true value of the correction factor according to:

$$f_{Marie}^{true} = \frac{14.41}{16} = 0.901$$

We can now use the non-integer value of K_{true}^{++} to get a good approximation of throughput in a zero-buffer queueing network. As proposed by Rall (1998, p. 126), the method of Marie is run twice using $\lceil K_{true}^{++} \rceil$ and $\lfloor K_{true}^{++} \rfloor$ as inputs and weighting the two results:

$$\lambda_{ZB}(K) \approx (\lceil K_{true}^{++} \rceil - K_{true}^{++}) \cdot \lambda_{Marie}(\lfloor K_{true}^{++} \rfloor)$$
$$+ (K_{true}^{++} - \lfloor K_{true}^{++} \rfloor) \cdot \lambda_{Marie}(\lceil K_{true}^{++} \rceil)$$

$$\lambda_{ZB}(20) = 2.72 \approx (15 - 14.41) \cdot \lambda_{Marie}(14) + (14.41 - 14) \cdot \lambda_{Marie}(15)$$
$$\lambda_{ZB}(20) = 2.72 \approx 0.59 \cdot 2.663 + 0.41 \cdot 2.792$$
$$\lambda_{ZB}(20) = 2.72 \approx 2.716$$

Note that this weighting procedure results in a better approximation than simply using an integer value and obtaining $\lambda_{Marie}(14) = 2.66$.

For f_{ZB} the procedure works accordingly with formula 5.15 changing to:

$$K_{true}^{+} = \frac{\lambda_{ZB}(K) - K_{x+1} \cdot \lambda_{UB}(K_x) + K_x \cdot \lambda_{UB}(K_{x+1})}{\lambda_{UB}(K_{x+1}) - \lambda_{UB}(K_x)} \tag{5.17}$$

and the f_{ZB} is calculated according to:

$$f_{ZB}^{true} = \frac{K_{true}^{+}}{K} \tag{5.18}$$

We ran this procedure for every single experiment of *experiment set 1* (see chapter 5.2.1) and obtained the values for f_{ZB}^{true} and f_{Marie}^{true}. In the following chapters, we will derive functional relations that estimate the values of f_{ZB}^{true} and f_{Marie}^{true} based on different network parameters.

5.3.3 Characteristic Network Parameters

Before focusing on the correction factors, we define some characteristic network parameters that will facilitate the derivation of functional relations between the correction factors and the input parameters.

Representative Network Variability for Zero-Buffer Networks

The integrated approach of Rall uses the representative network variability c_R^2, which weights the variabilities of each node to get one single representative value:

$$c_R^2 = \frac{\sum_{i=1}^{N} e_i \cdot \rho_i^2 \cdot 1/\sqrt{m_i} \cdot \overline{k_i}/B_i \cdot c_i^2}{\sum_{i=1}^{N} e_i \cdot \rho_i^2 \cdot 1/\sqrt{m_i} \cdot \overline{k_i}/B_i}$$

We pursue this idea but adjust the factor to the networks used in this work. First, as we only deal with single-server nodes, we can omit the term $1/\sqrt{m_i}$. Secondly, due to zero-buffers, the term $\overline{k_i}/B_i$ seems redundant as in zero-buffer networks the effect of this ratio is already incorporated in the utilization ρ_i. We also omit this term and will calculate the adjusted representative network variability for zero-buffer networks by:

$$c_{R,ZB}^2 = \frac{\sum_{i=1}^{N} e_i \cdot \rho_i^2 \cdot c_i^2}{\sum_{i=1}^{N} e_i \cdot \rho_i^2} \tag{5.19}$$

Gini Coefficient for Picker Distributions in Class-Based Storage

As class-based storage can have a large influence on the throughput in zero-buffer networks, we define a measure which contains an information on how pickers are distributed in the network. In detail, we want to know whether pickers are evenly distributed over the locations or if pickers tend to concentrate on selected areas of the network.

The Gini coefficient G measures the inequality of a population. A value of 0 expresses that each element of the population has an equal portion of an overall value and thus total equality. A value of 1 represents maximum inequality, thus one element of the population having the total overall value. In common applications, wealth or income are used as the overall value. The calculation of the Gini coefficient is based on the Lorenz curve, which represents the cumulative distribution function and thus the information how many bottom percent of population elements hold how many percent of the overall value (Fahrmeir et al. 2007, p. 78).

We adapt this concept to order picking systems and use the visit ratios e_s of all picking segments, i.e. $s = 1, 2, ..., \nu h$. Thus we figure out how many percent of picking segments hold how many percent of the sum of all visit ratios. In the following, let $e_{s,rel}$ denote the relative portion of a single visit ratio e_s compared to total visit ratios for all picking segments. They can be calculated according to:

$$e_{s,rel} = \frac{e_s}{\sum_{i=1}^{\nu h} e_i} \qquad \forall s = 1, 2, ..., \nu h$$

Next, we have to sort the values of $e_{s,rel}$ in ascending order and renumber them such that

$$e_{1,rel} \leq e_{2,rel} \leq e_{3,rel} \leq \cdots \leq e_{\nu h-1,rel} \leq e_{\nu h,rel}$$

The Gini coefficient can then be calculated by:

$$G = \frac{2 \cdot \sum_{s=1}^{\nu h} s \cdot e_{s,rel}}{\nu h \cdot \sum_{s=1}^{\nu h} e_{s,rel}} - \frac{\nu h + 1}{\nu h} \tag{5.20}$$

Finally, the Gini coefficient is normalized on the interval $[0, 1]$ by:

$$G^* = \frac{\nu h}{\nu h - 1} \cdot G \tag{5.21}$$

Taking the example in chapter 5.3.1 as a basis, we calculate $G^* = 0.01$ for random storage, $G^* = 0.16$ for medium-skewed class-based storage and $G^* = 0.37$ for high-skewed class-based storage.

5.3.4 Correction Factor f_{ZB}

From *experiment set 1* we calculate the values of f_{ZB}^{true} using formulas (5.17) and (5.18). With this, we basically answer the question "which number of customers K^+ would have led to the best approximation $\lambda_{UB}(K^+) \approx \lambda_{ZB}(K)$?". To eliminate any error from analytical approximations, e.g. the method of Marie, simulation was used to obtain λ_{UB}. Trends of f_{ZB}^{true} and an analysis on how different parameters influence the result shall be discussed first. We will then propose and validate a functional relation linking f_{ZB}^{true} to the picker density ω and the Gini co-efficient G^*, indicating how these pickers are distributed over the system.

The accuracy analysis of Rall's method (see chapter 5.2) has shown that congestion will be increased for storage location assignment policies leading to an uneven distribution of pickers to the segments. We therefore split the analysis in two parts and discuss random storage and class-based storage separately.

Random Storage: Trend Pattern for f_{ZB}^{true}

Figure 5.11 shows the scatter plot of the true correction factor f_{ZB}^{true} depending on the picker density ω for those experiment with random storage.

It appears as if the values for f_{ZB}^{true} follow a linear trend with a small negative slope. The plot also features a few stray values[6]. We use Pearson's r

$$r = \frac{\sum_{i=1}^{E}(x_i - \overline{x})(f_i - \overline{f})}{\sqrt{\sum_{i=1}^{E}(x_i - \overline{x})^2}\sqrt{\sum_{i=1}^{E}(f_i - \overline{f})^2}} \tag{5.22}$$

to calculate the correlation coefficient as $r = -0.708$. A simple linear regression would most likely result in a good approximation of f_{ZB}^{true}. However, we want to include *both* of the following thoughts:

---

[6]These can quickly be explained as follows: In networks with a high service time at the depot, it will act as a bottleneck for large ω. Even if the throughput loss caused by zero-buffer configurations is relatively small, the reduction in the number of customers from K to K^+ has to be quite big in order to leave the saturation level.

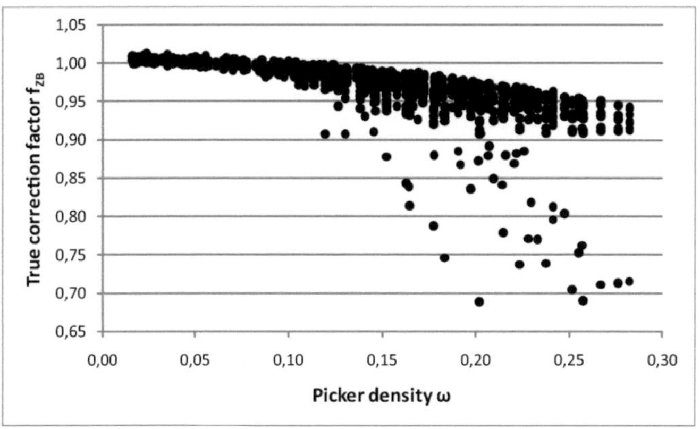

Figure 5.11: Trend for correction factor f_{ZB}^{true} - random storage

- Congestion stemming from level-2,3,...-blocking situations will be almost non-existent for very small ω, i.e. $f_{ZB} \approx 1$ for $\omega \approx 0$

- As ω increases, so will congestion and the influence of an additional picker for any ω_2 will be larger than the influence of an additional picker for any ω_1, if $\omega_2 > \omega_1$. In other words, relative congestion increase will be larger if the network is already fuller.

Random Storage: Functional Relation of f_{ZB} and Validation

Based on the trend pattern, we formulate the following simple equation:

$$f_{ZB}^{true} \approx f_{ZB} = 1 - \omega^2 \tag{5.23}$$

As $\omega < 1$ the quadratic influence will lead to a behavior of the trend of f_{ZB} as described above.

We apply formula 5.23 to those experiments with random storage and get a very good approximation of f_{ZB}^{true}. In figure 5.12 the approximation error ε between true values f_{ZB}^{true} and approximate values f_{ZB} is aggregated in deviation classes. For random storage, we can record the

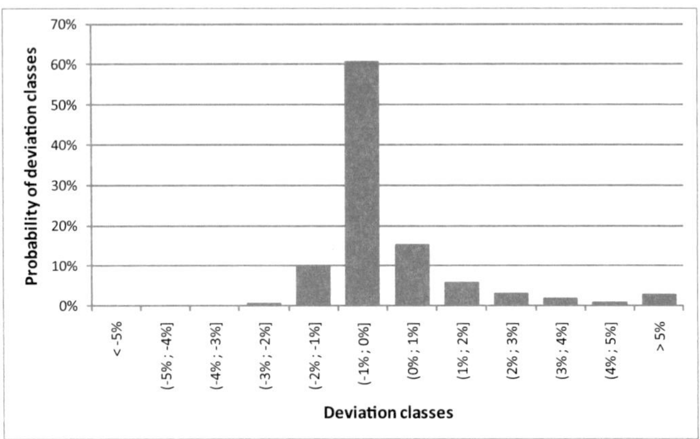

Figure 5.12: Deviation between f_{ZB}^{true} and f_{ZB} - random storage

following characteristics for ε:

- $P(\varepsilon \leq |5\%|) = 98.9\%$
- $P(\varepsilon \leq |10\%|) = 99.9\%$

Class-Based Storage: Trend Patterns for f_{ZB}^{true}

Figure 5.13 shows the scatter plot for only those experiments with a class-based storage policy. Results from experiments with a high-skewed class-based storage are shown in black dots. The correction factor tends to be smaller for high-skewed class-based storage. This seems logic, as we expect more congestion if more pickers meet in the aisles close to the depot and spend more time there. Nevertheless the trend of f_{ZB} is not unique. For the same value of ω, the range of values can differ noticeably. We can record values of f_{ZB} between 0.5 and 0.95 for $\omega \approx$ 0.2. Thus, in order to find a functional relation, we need to consider further dependencies between the input parameters. Considering figure 5.13 yields some general observations:

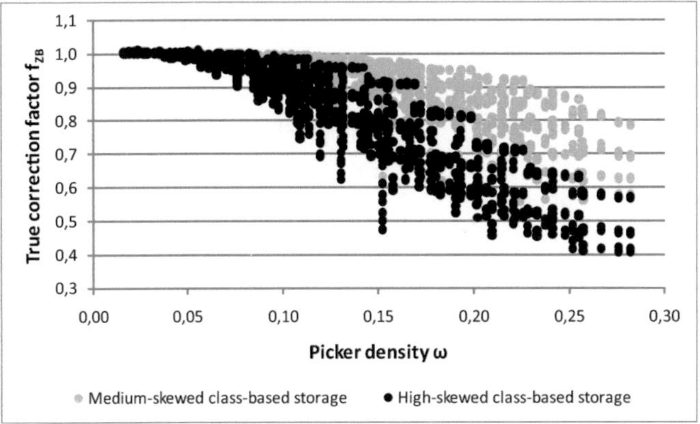

Figure 5.13: Trend for correction factor f_{ZB}^{true} - class-based storage

- For picker densities $\omega \leq 0.05$, the correction factor is very close to the value of 1, i.e. there is merely any influence of zero-buffers independent of the network parameters.

- Beginning at $\omega \approx 0.05$ there is an increasing influence. For the same value of ω, the influence is large for some network configurations and small for other network configurations. Thus the value of f_{ZB}^{true} depends on more parameters than just ω.

- The correction factor attains values as low as 0.4, showing that for some distinct scenarios, the influence of non-existing buffers will be quite negative in terms of throughput.

For a better understanding, we first test whether the relation between f_{ZB}^{true} and any input parameter x is of linear nature. After transforming all input parameters to cardinal scale[7], Person's r is calculated for all experiments. The three following values are of special interest:

$$r_{\omega,f} = -0.8 \qquad r_{c_{R,ZB}^2,f} = -0.14 \qquad r_{G^*,f} = -0.04$$

---

[7]e.g. storage location assignment policy: medium-skewed class-based storage $= 1$; high-skewed class-based storage $= 2$

None of the relations is of significant linear nature. Still, the relation between the correction factor and the picker density ω is indisputable as we tend to have a smaller correction factor for increasing ω. Surprisingly, it seems virtually impossible to find any useful functional relation between $c_{R,ZB}^2$ and f_{ZB}^{true}. The same holds true when relating G^* and f_{ZB}^{true}.

Because a one-dimensional relation is not sufficient, we try to find a two-dimensional relation, linking ω with another parameter:

- f_{ZB}^{true} as a function of ω *and* $c_{R,ZB}^2$
- f_{ZB}^{true} as a function of ω *and* G^*

The combination of picker density and Gini coefficient results in a comparatively smooth running scatter plot as illustrated in figure 5.14.

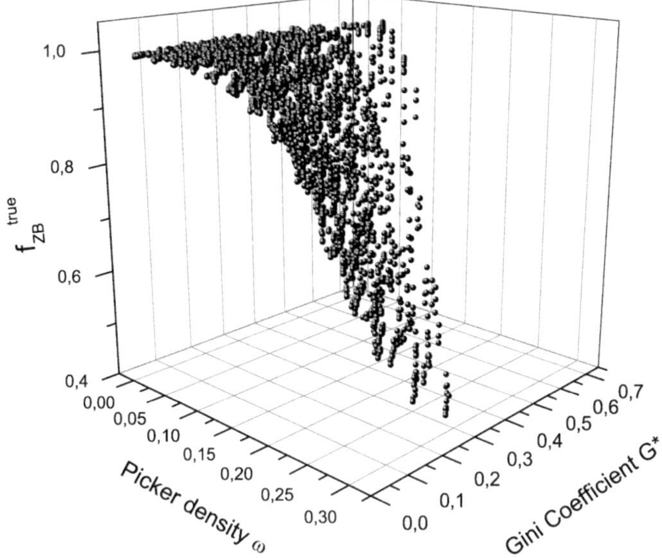

Figure 5.14: Scatter plot for f_{ZB}^{true} as a function of ω and G^*

There are only very few runaway values and a particular relation between the two predictors (ω and G^*) and the correction factor f_{ZB} seems existent. We will therefore use a nonlinear regression model with two predictor variables to determine the relation.

Class-Based Storage: Functional Relation for f_{ZB} and Validation

We refer to the regression analysis as *surface fitting* because we have two predictor variables and thus try to fit a surface to our calculated values f_{ZB}^{true}.

We preselect several possible fit-functions and perform a surface fitting iteration[8]. This includes adjusting the parameters of the respective fit-functions in order to minimize the sum of squared residuals between calculated values (f_{ZB}^{true}) and approximated values (f_{ZB}). The fitness of each fit-function will be evaluated by the following three criteria:

- Evaluation of the coefficient of determination R^2

- Evaluation of $\chi^2_{reduced}$

- Graphical evaluation of the percentage deviation between f_{ZB}^{true} and f_{ZB}.

For E experiments, the coefficient of determination R^2, also called Goodness-of-Fit-Index, is calculated by:

$$R^2 = 1 - \frac{\sum_{i=1}^{E}(f_{ZB,i}^{true} - f_{ZB,i})^2}{\sum_{i=1}^{E}(f_{ZB,i}^{true} - \overline{f_{ZB}})^2} \tag{5.24}$$

Note that the numerator corresponds to the portion of variation that is not explained by the regression model. Subtracting the numerator from the denominator will correspond to the portion of variation that is explained by the regression model. A value of $R^2 = 1$ represents a perfect fit.

Let df be the degrees of freedom, then $\chi^2_{reduced}$ is calculated by:

$$\chi^2_{reduced} = \frac{\chi^2}{df} = \frac{\sum_{i=1}^{E}(f_{ZB,i}^{true} - f_{ZB,i})^2}{df} \tag{5.25}$$

According to Backhaus et al. (2003, p. 373) a model represents a good fit, as long as $\chi^2_{reduced} \leq 2.5$. However, we need to be careful as the value

---

[8]We used the software tool Origin-Pro version 8

of $\chi^2_{reduced}$ might lead to false interpretations if underlying variables are not normally distributed.

Table 5.3 shows the fit-functions we tested and the respective results of the surface fitting iteration (see appendix B for a detailed structure of the fit-function formulas).

Fit-Function	x	y	R^2	$\chi^2_{reduced}$	Formula (appendix B)
Cosine2D	ω	G^*	0.932	0.00124	B.1
Plane	ω	G^*	0.757	0.00439	B.3
Power2D	ω	G^*	0.950	0.00091	B.5
Gauss2D	ω	G^*	0.949	0.00092	B.7
Poly2D	ω	G^*	0.953	0.00085	B.9
Cosine2D	ω	$c^2_{R,ZB}$	0.683	0.00573	B.2
Plane	ω	$c^2_{R,ZB}$	0.651	0.00632	B.4
Power2D	ω	$c^2_{R,ZB}$	0.683	0.00575	B.6
Gauss2D	ω	$c^2_{R,ZB}$	0.683	0.00573	B.8
Poly2D	ω	$c^2_{R,ZB}$	0.681	0.00577	B.10

Table 5.3: Goodness-Of-Fit for different fit-functions of f_{ZB}

The results indicate that the combination of ω and G^* enables a better approximation compared to the combination of ω and $c^2_{R,ZB}$. Most functions using the Gini coefficient result in a large coefficient of determination R^2 and small values of $\chi^2_{reduced}$.

We select the *Poly2D* fit-function to approximate the values f^{true}_{ZB}. A Poly2D function basically has the following structure:

$$z = g(x, y) = z_0 + A \cdot x + B \cdot y + C \cdot x^2 + D \cdot y^2 + F \cdot x \cdot y$$

To run the surface fitting iteration procedure for those experiments with class-based storage, we use f^{true}_{ZB} as the dependent variable, while ω and G^* act as the independent variables, i.e. predictors. The iteration results in the following parameter values:

$z_0 = 0.94 \qquad A = 1.82 \qquad B = 0.25 \qquad C = 9 \qquad D = 0.001 \qquad F = 6$

For simplification, we omit the term $0.001 \cdot (G^*)^2$ and get the following equation to calculate the correction factor f_{ZB} for class-based storage:

$$f_{ZB}^{true} \approx f_{ZB} = 0.94 + 1.82\omega + 0.25G^* + 9\omega^2 + 6\omega G^* \qquad (5.26)$$

Finally, formula (5.26) is applied to those experiments with class-based storage and we obtain a distribution of deviation classes as shown in figure 5.15.

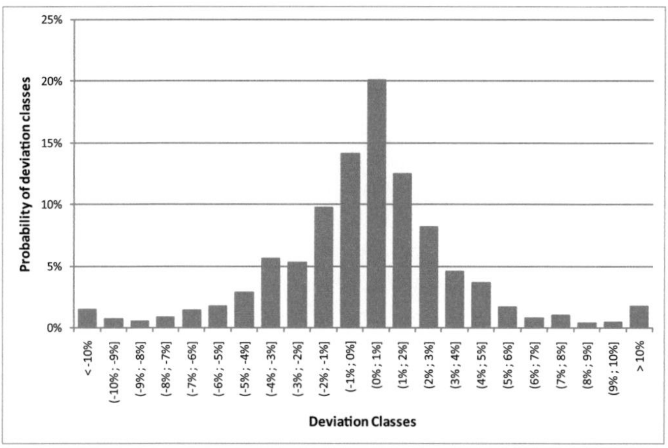

Figure 5.15: Deviation between f_{ZB}^{true} and f_{ZB} - class-based storage

Overall 86% of true correction factors were approximated within a $|5\%|$-error range and 97% of true correction factors were approximated within a $|10\%|$-error range. These values are not as good as those for random storage. However they are of sufficient accuracy because the error in the approximation of the correction factor will most likely not result in an error of the same size when approximating throughput.

Summary of f_{ZB}

We briefly summarize the formulas to calculate the correction factor f_{ZB}:

$$f_{ZB} = \begin{cases} 1 - \omega^2, & \text{for random} \\ 0.94 + 1.82\omega + 0.25G^* + 9\omega^2 + 6\omega G^*, & \text{for class-based} \end{cases} \quad (5.27)$$

The overall accuracy of these formulas for both storage policies can be characterized by the approximation error ε between f_{ZB}^{true} and f_{ZB}:

- $P(\varepsilon \leq |5\%|) = 90\%$
- $P(\varepsilon \leq |10\%|) = 97\%$

5.3.5 Correction Factor f_{Marie}

Again we use *experiment set 1* to calculate the values of f_{Marie}^{true} using formulas (5.15) and (5.16). This enables us to answer the question "which number of customers K^{++} would have led to the best approximation $\lambda_{Marie}(K^{++}) \approx \lambda_{UB}(K^+)$?". Simulation was used to obtain λ_{UB} and λ_{Marie} was calculated analytically. We learned earlier (see chapter 5.2.5) that Marie's method will over- or underestimate throughput and that the error strongly depends on the variabilities in the network.

Trend Patterns for f_{Marie}^{true}

Figure 5.16 shows the scatter plot for the values of f_{Marie}^{true} as a function of picker density ω. Network variabilities $c_{R,ZB}^2 \leq 1$ are indicated by grey dots. The following interpretations can be drawn from the graph:

- The correction factor quickly drifts away from the level $f_{Marie}^{true} \approx 1$. Thus we have a considerable deviation even for small values of picker density ω

- For the same values of ω, there will be different values of f_{Marie}^{true}. Thus the value of f_{Marie}^{true} depends on more than just ω. Analogous, for the same value of $c_{R,ZB}^2$, there will be different values of f_{Marie}^{true}. We therefore need to consider more than just $c_{R,ZB}^2$ in a functional relation

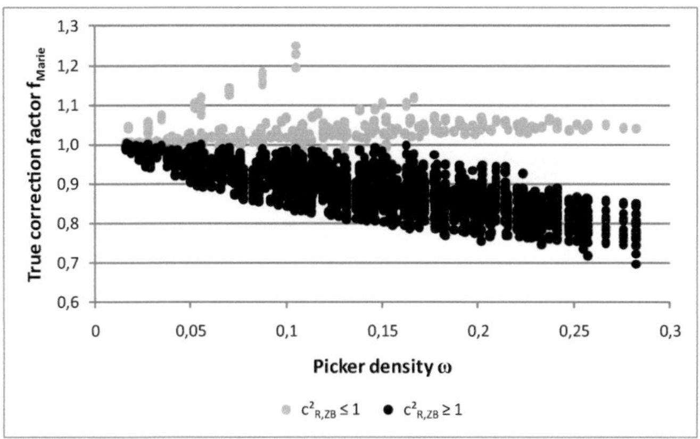

Figure 5.16: Scatter plot for f_{Marie}^{true} as a function of ω

As for f_{ZB} we try to find out whether the relation is of linear structure and calculate Person's r. The three following results are of special interest:

$$r_{\omega,f} = -0.64 \qquad r_{c_{R,ZB}^2,f} = -0.73 \qquad r_{G^*,f} = 0.25$$

Again, the relations are not of significant linear nature. We further analyzed the experimental data trying to find a relation between the parameters with the biggest effect, namely ω and $c_{R,ZB}^2$. In most cases, the trend of f_{Marie}^{true} over ω is increasing for $c_{R,ZB}^2 \leq 1$ and increasing stronger for decreasing $c_{R,ZB}^2$. Likewise, the trend of f_{Marie}^{true} over ω is decreasing for $c_{R,ZB}^2 > 1$ and decreasing stronger for increasing $c_{R,ZB}^2$. Figure 5.17 includes the trends from some arbitrary network configurations, supporting this statement[9]. We can identify a family of curves, basically relating ω to f_{Marie}^{true} while the position and gradient of the curve is determined by the network variability $c_{R,ZB}^2$.

---

[9]Note that not all trends have values for $\omega \geq 0.2$. This is because based on the network configuration, $\omega_{seg,max}$ will lead to different ω_{max}.

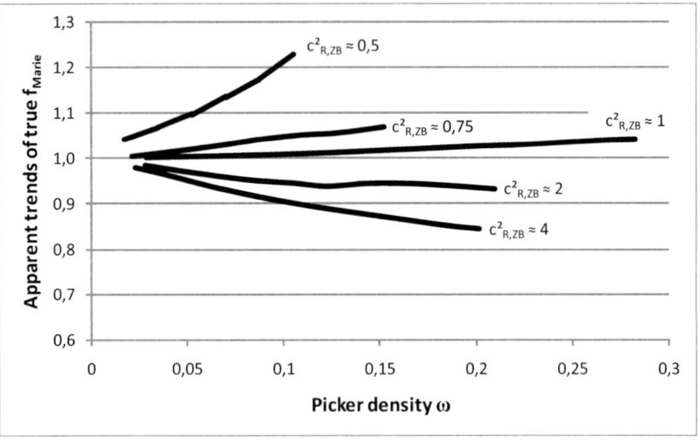

Figure 5.17: Trends for f_{Marie}^{true} for selected $c_{R,ZB}^2$

Functional Relation for f_{Marie} and Validation

As the trends in figure 5.17 appear to be linear, we seek to approximate the trend for each specification of the network variability $c_{R,spec}^2$ by a simple straight line depending on ω. Experiments are grouped by ascending network variabilities $c_{R,ZB}^2$ and we calculate the respective parameters of the line by linear regression. As almost 75% of groups resulted in Pearson's $r \geq 0.95$, we tried to approximate the line parameters by some functional relation with other network parameters. This basically works, however accuracy is reduced by the fact that the straight lines react quite sensitive on the choice of parameters, thus propagating errors would be incorporated in the procedure.

We therefore use a power function to approximate the values of f_{Marie}^{true} and thus the family of curves. We choose it to be of form:

$$f_{Marie}^{true} \approx f_{Marie} = c_{R,ZB}^2 {}^{(-A \cdot \omega + B)}$$

This choice is based on the following thoughts:

156

- For $c_{R,ZB}^2 \approx 1$, the function will be very close to a straight line. This is plausible as the method of Marie produces good approximations if the network is based on exponential distributions.

- For $c_{R,ZB}^2 < 1$, the negative term $-A \cdot \omega$ will likely lead to a value $f_{Marie} > 1$. The adjustment factor B can be used to fine-tune this behavior.

- For $c_{R,ZB}^2 > 1$, the negative term $-A \cdot \omega$ will likely lead to a value $f_{Marie} < 1$. Again, the adjustment factor B can be used to fine-tune this behavior.

We performed a function-fit iteration procedure to get an estimation for the parameters A and B, such that the sum of squared residuals between calculated values (f_{Marie}^{true}) and approximated values (f_{Marie}) is minimized. The procedure successfully resulted in the following parameters:

$$A = \frac{2}{3} \qquad B = \frac{15}{1000}$$

and thus we can formulate a function relating the correction factor to the input parameters of the network:

$$f_{Marie}^{true} \approx f_{Marie} = c_{R,ZB}^2 {}^{\left(-\frac{2}{3} \cdot \omega - \frac{15}{1000}\right)} \tag{5.28}$$

To confirm validity of this estimator function, we first calculate the coefficient of determination R^2 (formula 5.24) as 0.895 and the $\chi_{reduced}^2$-value (formula 5.25) is calculated as 0.0005604 . The approximation does not reach the accuracy of f_{ZB}, nevertheless an $R^2 \approx 0.9$ is a good indicator that the approximation is working well.

Secondly, we use formula (5.28) for *experiment set 1* and obtain good approximation of f_{Marie}^{true}. Figure 5.18 includes the deviation between true values f_{Marie}^{true} and approximate values f_{Marie}. We can also consider the overall deviation classification according to:

- $P(\varepsilon \leq |5\%|) = 89\%$
- $P(\varepsilon \leq |10\%|) = 99.9\%$

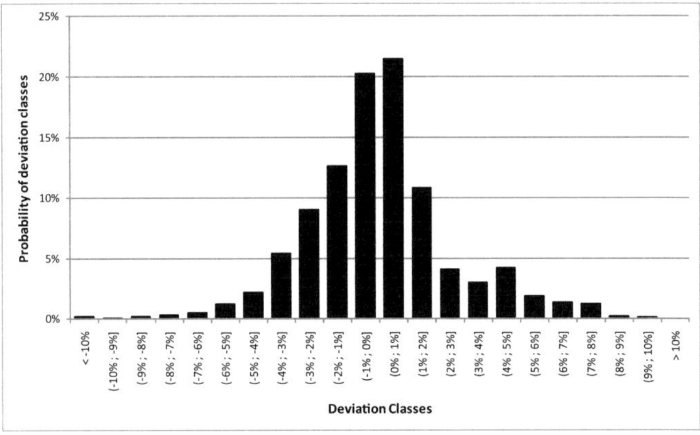

Figure 5.18: Deviation between f_{Marie}^{true} and f_{Marie}

5.4 Chapter Conclusion

Let us summarize the steps of our new integrated approach to calculate the throughput in a closed zero-buffer queueing network with K customers, $\lambda_{ZB}(K)$ and with it the throughput of a narrow-aisle manual order picking system with K pickers, $\lambda_{OPS}(K)$:

- Transform the parameters of an order picking system into the parameters of a closed zero-buffer queueing network. In particular, we need to define:
    - the resources of a system (aisle, cross aisle, depot), i.e. the number of nodes N, according to chapter 4.3
    - the visit ratios, i.e. the parameters defining the routing policy according to chapters 4.5 (Random storage) and 4.6 (Class-Based storage)
    - the service times and variabilities according to chapter 4.7
- Calculate a product-form algorithm to obtain the utilization ρ_i for each node

- Calculate the representative network variability $c_{R,ZB}^2$ with formula (5.19) as well as the Gini coefficient with formula (5.21)
- Calculate the correction factor f_{ZB} with formula (5.27)
- Calculate the correction factor f_{Marie} with formula (5.28)
- If $K \geq 3$, calculate the adjusted number of customers K^{++} according to:

$$K^{++} = K \cdot f_{ZB} \cdot f_{Marie}$$

If $K < 3$, calculate K^{++} according to:

$$K^{++} = K \cdot f_{Marie}$$

- Use K^{++} as an input in Marie's method. As K^{++} might be a non-integer value, Marie's algorithm will be run twice using $\lfloor K^{++} \rfloor$ and $(\lfloor K^{++} \rfloor + 1)$ respectively. The final result is obtained by weighting the results of the two runs according to:

$$\lambda_{ZB}(K) \approx \lambda_{Marie}(K^{++}) = a \cdot \lambda_{Marie}(\lfloor K^{++} \rfloor) + b \cdot \lambda_{Marie}(\lfloor K^{++} \rfloor + 1)$$

where the parameters a and b are defined as follows:

$$a = \lfloor K^{++} \rfloor + 1 - K^{++} \qquad b = K^{++} - \lfloor K^{++} \rfloor$$

- Calculate the throughput of the order picking system (order lines per period of time) by multiplying the result of the new integrated approach with the number of picks per order:

$$\lambda_{OPS}(K) \approx \lambda_{ZB}(K) \cdot n$$

6 Validation and Application

We will focus on the validation of the new integrated approach in chapter 6.1. The presented procedure will be tested by using a wide range of order picking system configurations. The same parameters are then used in chapter 6.2 to quantify the effects of congestion in narrow-aisle systems.

6.1 Validation

We will validate our new integrated approach by analyzing its accuracy for a large number of experiments. We calculate the approximation error by comparing the throughput of our analytical approach to the throughput of simulation:

$$\varepsilon = 100 \cdot \frac{\lambda_{\text{analytic}} - \lambda_{\text{simulative}}}{\lambda_{\text{simulative}}}$$

The mean approximation error $\bar{\varepsilon}$ over all E experiments is calculated as the average value of all approximation errors where we do not differentiate between positive and negative errors:

$$\bar{\varepsilon} = 100 \cdot \frac{\sum_{i=1}^{E} \left| \frac{\lambda_{\text{analytic}} - \lambda_{\text{simulative}}}{\lambda_{\text{simulative}}} \right|}{E}$$

In addition to the single values we also present some information on the distribution of the errors ε by indicating its minimum and maximum values as well as calculating the quartiles[1].

We will first reconsider *experiment set 1* and afterwards extend this set in terms of picking time variability to determine to which degree this

---

[1] For instance, if $Q_3 = 1\%$ then 75% of respective experiments have an error $\varepsilon \leq 1\%$.

single parameter will influence the accuracy. Thereafter we will apply the approach to *experiment set 2*, a new set of experiments with input parameters strongly differing from those of set 1. We do this in order to confirm that the approach is working for parameters *not* used in the derivation process of the correction factors f_{ZB} and f_{Marie}. Finally, we show to what extend our new approach improves the approximation accuracy of Rall's integrated approach (1998).

6.1.1 Experiment Set 1

We used *experiment set 1* to determine how accurate the approach of Rall was working for queueing models of order picking systems (chapter 5.2) and for the derivation of the correction factors f_{ZB} and f_{Marie} (chapters 5.3.4 and 5.3.5). The set includes 4860 experiments and the parameters are given on page 121.

Overall Accuracy

The histogram of deviations between the new integrated approach and simulation is given in figure 6.1. The probability mass function shows that the errors are rather normally distributed around the value of 0%, i.e. there is no distinct tendency to under- or overestimate throughput in an order picking system. Overall, the mean error $\bar{\varepsilon}$ is 2% and we observe the following:

- $P(\varepsilon \leq |5\%|) = 93\%$
- $P(\varepsilon \leq |10\%|) = 99\%$
- $Min = -29.4\%$; $Q_1 = -1.1\%$; $Q_2 = 0.3\%$; $Q_3 = 1.7\%$; $Max = 12.1\%$

Thus for *experiment set 1* the overall performance of the new approach is very good. Still, in order to analyze the few runaway values in more detail, we will consider the accuracy based on different input parameters.

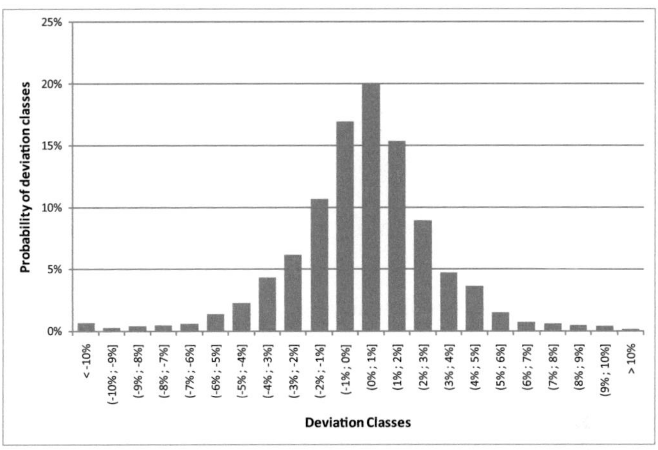

Figure 6.1: Deviation between new approach and simulation - *experiment set 1*

Accuracy Based on Storage Location Assignment Policy

For a better overview, we first classify the results by their storage policy. Table 6.1 indicates some information on the accuracy.

SLAP	$P(\varepsilon \leq |5\%|)$	$P(\varepsilon \leq |10\%|)$	Min	Q_1	Q_2	Q_3	Max
Random	100	100	-3.8	-0.6	0.3	1	3.4
Medium-Skewed	95	99	-12.7	-1.2	0.2	1.8	6.1
High-Skewed	82	98	-29.4	-2.6	0.5	2.9	12.1
	all values in %						

Table 6.1: Accuracy characteristics of the new approach for different storage policies - *experiment set 1*

The new approach works very well for random storage policy as all experiments had an error of less than 5%. For a medium-skewed class-based storage, we also get very good results with very moderate and few runaway values. For high-skewed class-based storage we still get a good accuracy, however with some runaway values. These are mostly underestimations of throughput stemming from experiments with large ω_{seg} and small pick densities.

For this parameter combination, the correction factor f_{ZB} can sometimes be too small. The Gini coefficient tends to be high for small pick density, because visit ratios in those aisles farthest from the depot will be very small. Both high ω_{seg} and high G^* will results in a lower f_{ZB}.

Accuracy Based on Picker Density

Our integrated approach works best for smaller ω_{seg} and accuracy slightly decreases with increasing ω_{seg}. Table 6.2 includes quartiles and minimum as well as maximum values of the error. For certain experiments, large ω_{seg} in combination with other parameters will imply certain blocking situations. They will be covered correctly if the empirical correction factor f_{ZB} works accurately. For some parameter configurations, the adjustment of the number of pickers will be too big or too small. Note that a few errors can also be traced back to the correction factor f_{Marie} with f_{ZB} working very well. However in the majority of cases, f_{ZB} is the root cause for the error (we also refer to the validation of the correction factors in chapters 5.3.4 and 5.3.5).

Accuracy Based on Warehouse Shape Factor (WSF)

We also analyze the accuracy of the new approach when classified by the warehouse shape factor. Our approach has the highest accuracy for smaller WSF, i.e. when the order picking system tends to have many short aisles instead of few long aisles. Table 6.3 gives the characteristic accuracy values.

Accuracy Based on Remaining Parameters

For the remaining parameters we give the following brief summary:

- Number of aisles ν: although we cannot identify a clear trend, the accuracy is highest for medium number of aisles and most runaway

ω_{seg}	$P(\varepsilon \leq |5\%|)$	$P(\varepsilon \leq |10\%|)$	Min	Q_1	Q_2	Q_3	Max
0.0333	100	100	-2.6	-0.6	0.2	1.5	4.9
0.0667	99	100	-3.4	-0.2	1	2.3	5.5
0.1	97	100	-3.4	-0.2	1.2	2.6	5.9
0.1333	94	100	-5.8	-0.5	1.2	2.5	6.2
0.1667	92	100	-6.6	-1	0.9	2.3	7.8
0.2	89	100	-8	-1.2	0.5	1.9	10.4
0.2333	88	99	-8	-1.2	0.3	1.3	11
0.2667	91	100	-9.4	-1.4	-0.1	0.7	9.2
0.3	89	99	-11	-2.6	-1	0.5	7.5
0.3333	78	93	-29.4	-4.3	-2	0.4	12.1
	all values in %						

Table 6.2: Accuracy characteristics of the new approach for different ω_{seg} - *experiment set 1*

WSF	$P(\varepsilon \leq |5\%|)$	$P(\varepsilon \leq |10\%|)$	Min	Q_1	Q_2	Q_3	Max
0.333	95	100	-9.4	-1.1	0.2	1.1	12.1
0.667	92	100	-22.3	-1.3	0.2	1.7	11
1	89	98	-29.4	-1.2	0.6	2.3	10.3
	all values in %						

Table 6.3: Accuracy characteristics of the new approach for different WSF - *experiment set 1*

values were recorded for large number of aisles. This strongly correlates to the number of nodes in the network: accuracy tends to be highest for smaller networks (note that in this context "small" networks still include up to 300 nodes).

- Pick density: accuracy is higher for larger pick densities and few runaway values were recorded for small pick densities (in combination with high-skewed class-based storage as described above).

- Representative network variability $c_{R,ZB}^2$: If the approximation error is noteworthy at all, it will likely be an underestimation for $c_{R,ZB}^2 \leq 1.5$ and an overestimation for $c_{R,ZB}^2 > 1.5$. However, there are not clearly identifiable trends.

- Service time $t_{s,Depot}$: the accuracy is very good for all three parameter scenarios thus the service time at the depot does not seem to have a big influence. However, if there are many pickers in the system (bigger ω_{seg}), large service times at the depot slightly improve the accuracy. This is because the depot then will be the obvious bottleneck of the order picking system and there will always be pickers queueing in front of the depot. In such cases, the congestion within the aisles tends to be irrelevant. Furthermore, any approximation error of f_{ZB} and/or f_{Marie} will be with very few or no influence as throughput has already reached the degree of saturation and reacts very insensitive when the number of customers is adjusted.

Table 6.4 includes the characteristic accuracy values for the remaining parameters.

6.1.2 Extended Experiment Set 1

In the extension of *experiment set 1*, we will get an idea to what extent the picking time variability c_{Pick}^2 will influence the accuracy of our new integrated approach.

ν	6	8	10	12	14	16
$P(\varepsilon \leq |5\%|)$	90.5	93.8	95	93.9	90.2	88.7
$P(\varepsilon \leq |10\%|)$	99	99.8	99.3	99.3	99.1	98.7
Pick Density	0.05	0.1	0.2			
$P(\varepsilon \leq |5\%|)$	87.1	92	95.9			
$P(\varepsilon \leq |10\%|)$	97.6	99.8	100			
$c^2_{R,ZB}$	≤ 1.5	> 1.5				
$P(\varepsilon \leq |5\%|)$	93.3	91.6				
$P(\varepsilon \leq |10\%|)$	100	98.9				
$t_{s,Depot}$	1	5	15			
$P(\varepsilon \leq |5\%|)$	91	91.1	93.8			
$P(\varepsilon \leq |10\%|)$	98.8	99.2	99.6			
	all values in %					

Table 6.4: Accuracy characteristics of the new approach for several input parameters - *experiment set 1*

Overall Accuracy

The extended set 1 includes 9720 experiments and the histogram of deviations between the new integrated approach and simulation is given in figure 6.2. In summary, the mean error $\bar{\varepsilon}$ is 2.1% for $c^2_{Pick} = 0.25$ and $\bar{\varepsilon}$ is 2.5% for $c^2_{Pick} = 0.75$. Table 6.5 completes the key accuracy characteristics. We can summarize that the *extended experiment set 1* also does perform very good as more than 90% of all experiments have an approximation error of less than $|5\%|$. Except for a few runaway values, the approximation is always within a $|10\%|$-error range. Consideration of all experiments from set 1 (i.e. $c^2_{Pick} = 0.25; 0.5; 0.75$ with 14040 experiments) reveals that the new integrated approach basically calculates results of very good accuracy.

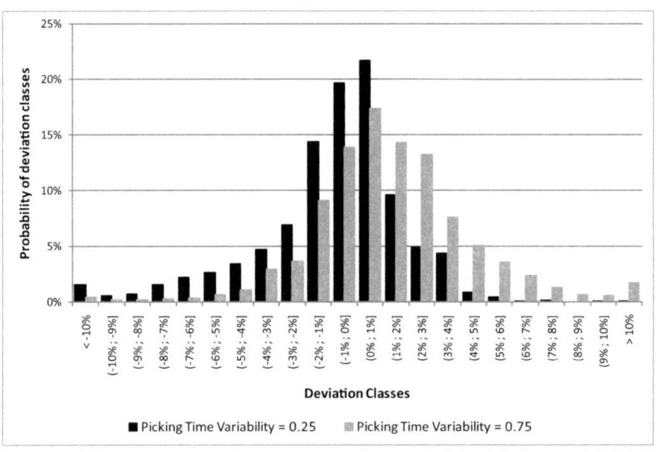

Figure 6.2: Deviation between new approach and simulation - *extended experiment set 1*

c^2_{Pick}	$P(\varepsilon \leq |5\%|)$	$P(\varepsilon \leq |10\%|)$	Min	Q_1	Q_2	Q_3	Max
0.5	93	99	-29.4	-1.1	0.3	1.7	12.1
0.25	90	98	-31.8	-1.9	-0.4	0.7	10.5
0.75	88	98	-27.7	-0.5	1	2.8	14.3
All	90	99	-31.8	-1.2	0.3	1.8	14.3
	all values in %						

Table 6.5: Accuracy characteristics of the new approach - *extended experiment set 1*

Accuracy Based on Storage Location Assignment Policy

As for the original set 1, the accuracy is strongly influenced by the storage policy. For random storage, the approach again works very well as all experiments have errors smaller than $|5\%|$. The approach also works good for medium-skewed class-based storage and we experience some larger errors for high-skewed class-based storage. For $c^2_{Pick} = 0.25$ those tend to be underestimations stemming from the fact that the correction factor f_{ZB} will "expect" more blocking situations as there will actually be. Vice versa, for $c^2_{Pick} = 0.75$ throughput tends to be overestimated - if there is an error - as there is more instability in the network than actually included in f_{ZB}. As before, this error occurs if the respective storage location assignment policies are combined with high picker densities ω_{seg}. Table 6.6 includes the characteristic accuracy values.

SLAP	$P(\varepsilon \leq |5\%|)$	$P(\varepsilon \leq |10\%|)$	Min	Q_1	Q_2	Q_3	Max
$c^2_{Pick} = 0.25$							
Random	100	100	-4.5	-0.9	0	0.6	3.2
Medium-Skewed	91	99	-13.9	-2	-0.5	0.9	4.8
High-Skewed	80	97	-31.8	-4	-1.2	1.2	10.5
$c^2_{Pick} = 0.75$							
Random	100	100	-3.4	-0.4	0.6	1.7	4.3
Medium-Skewed	91	100	-12.5	-0.5	1	2.8	7.3
High-Skewed	72	94	-27.7	-0.5	2.4	4.9	14.3
	all values in %						

Table 6.6: Accuracy characteristics of the new approach for different storage policies - *extended experiment set 1*

Accuracy Based on Picker Density

Again it is rather obvious that accuracy tends to be best for small ω_{seg} and decreases for larger ω_{seg}. Table 6.7 shows the characteristic data. For small ω_{seg}, the results for $c^2_{Pick} = 0.25$ seem to be a little better. Vice versa, for larger ω_{seg}, the results for $c^2_{Pick} = 0.75$ have a better approximation quality.

ω_{seg}	$P(\varepsilon \leq \|5\%\|)$	$P(\varepsilon \leq \|10\%\|)$	Min	Q_1	Q_2	Q_3	Max
$c_{Pick}^2 = 0.25$							
0.0333	100	100	-2.3	-0.7	0.1	1.2	4.9
0.0667	100	100	-2.5	-0.2	0.6	1.5	3.9
0.1	100	100	-3.4	-0.4	0.6	1.6	4.8
0.1333	97.9	100	-5.6	-0.9	0.3	1.4	4.7
0.1667	91.8	100	-8	-1.4	-0.1	0.8	4.2
0.2	87.8	100	-8.3	-1.7	-0.3	0.6	6.2
0.2333	88	100	-9.7	-2.3	-0.7	0.1	5.3
0.2667	89.9	99.8	-10.8	-2.9	-1.5	-0.2	7.4
0.3	80.2	96.4	-12.9	-4.3	-2.2	-0.8	7.9
0.3333	63.9	87	-31.8	-6.6	-3.3	-1.1	10.5
$c_{Pick}^2 = 0.75$							
0.0333	98.1	100	-2.7	-0.5	0.5	1.9	5.6
0.0667	95.5	100	-3.2	0.2	1.5	2.9	7.1
0.1	88.9	100	-3.2	0.2	1.9	3.6	7.8
0.1333	85	100	-2.9	-0.1	2	3.7	8.8
0.1667	83.7	99	-4.9	-0.5	1.7	3.4	10.8
0.2	87.4	96.1	-4.9	-0.2	1.4	3.1	13.7
0.2333	86.4	92.8	-5.5	-0.1	1.3	3.0	14.3
0.2667	85.7	96.6	-8	-0.4	0.9	2.5	13
0.3	84.6	99.5	-10.3	-1.3	0.2	1.6	11.1
0.3333	83.8	94.8	-27.7	-2.9	-0.9	1	14
	all values in %						

Table 6.7: Accuracy characteristics of the new approach for different ω_{seg} - *extended experiment set 1*

Accuracy Based on Remaining Parameters

- Number of aisles ν: for $c_{Pick}^2 = 0.25$, we cannot identify a clear trend to what extent the number of aisles influences the accuracy of the new approach. There is a slight tendency for more underestimation if ν is decreasing. For $c_{Pick}^2 = 0.75$, there is a trend, as accuracy decreases for increasing number of aisles.

- Warehouse Shape Factor WSF: for both picking time variabilities the new integrated approach works best for a small warehouse shape factor $WSF = 0.33$ and there are a some deviations larger than $|5\%|$ for $WSF = 1$, namely underestimations for $c_{Pick}^2 = 0.25$ and overestimations for $c_{Pick}^2 = 0.75$.

- Pick Density: for $c_{Pick}^2 = 0.25$, accuracy tends to be slightly better for medium pick densities. For $c_{Pick}^2 = 0.75$, accuracy is better for larger pick densities.

- Representative network variability $c_{R,ZB}^2$: for $c_{Pick}^2 = 0.25$, the approach tends to result in a slight underestimation for experiments with $c_{R,ZB}^2 \leq 1.5$. For larger $c_{R,ZB}^2$, the approximation is very good. This observation is contrary for $c_{Pick}^2 = 0.75$ as accuracy will be very good for $c_{R,ZB}^2 \leq 1.5$. For larger values of $c_{R,ZB}^2$, there will be a few overestimations of throughput. Still, overall accuracy is very good.

- Service time $t_{s,Depot}$: for both picking time variabilities, the service time at the depot does not seem to have a big influence.

Table 6.8 includes the characteristic accuracy values.

6.1.3 Experiment Set 2

Another validation run was performed with *experiment set 2*, which differs from set 1 in a way that some parameters were chosen to have different values. Some parameters, e.g. ν and the storage policy, remain unchanged. The purpose of this analysis is to show that the new integrated approach also produces good approximations for input parameters not used in the process of deriving the correction factors.

Parameters	$c^2_{Pick} = 0.25$		$c^2_{Pick} = 0.75$	
	$P(\varepsilon \leq |5\%|)$	$P(\varepsilon \leq |10\%|)$	$P(\varepsilon \leq |5\%|)$	$P(\varepsilon \leq |10\%|)$
Number of aisles ν				
6	88.1	98.4	96.2	99.3
8	87.8	99.3	96.5	99.9
10	92.1	99.4	90.7	98.2
12	91.8	98.6	86.7	98.4
14	91.5	97.9	81.1	96.6
16	90.7	97.1	76.4	95.2
Warehouse Shape Factor - WSF				
0.333	95	99.9	95.7	99.4
0.667	88.4	98.9	87.8	98.2
1	87.7	96.6	81	96.3
Pick Density				
0.05	90	96.5	82	96.3
0.1	93.5	98.9	86.7	97
0.2	87.9	99.6	94	100
Representative network variability $c^2_{R,ZB}$				
≤ 1.5	84.8	99.7	93.7	99.9
> 1.5	92.7	97.9	86.4	97.4
Service Time depot $t_{s,Depot}$				
1	90.9	98.5	86	97.3
5	90.4	98.3	86.7	97.4
15	89.7	98.6	91.4	99
	all values in %			

Table 6.8: Accuracy characteristics of the new approach for several input parameters - *extended experiment set 1*

Experiment Design

The following gives a brief overview on the parameters that were used:

- Parameters changed compared to experiment set 1
    - Warehouse shape factor WSF: $\frac{1}{2}$, $\frac{3}{4}$ (the shape factor is defined as the ratio $\frac{L}{W}$)
    - Pick Density: $\frac{3}{40}$, $\frac{3}{20}$, $\frac{1}{4}$ (the pick density is defined as the ratio $\frac{n}{\nu \cdot h}$)
    - Picker Density (based on segments) ω_{seg}: $\frac{3}{50}$, $\frac{3}{25}$, $\frac{9}{50}$, $\frac{6}{25}$, $\frac{3}{10}$
- Parameters unchanged compared to experiment set 1
    - Number of aisles ν: 6, 8, 10, 12, 14, 16
    - Storage Location Assignment Policy: Random, medium-skewed class-based, high-skewed class-based. We assume both class-based policies have three classes with the A-class occupying 20% of the segments, the B-class occupying 30% and the C-class occupying 50% of the segments. For medium-skewed, the picks will be distributed on these classes by 50/30/20%, i.e. 50% of the picks will be located in 20% of the segments. For high-skewed, this distribution will be 80/15/5%.
    - Service time $t_{s,Depot}$: 1, 15 (seconds)
    - Variability c^2_{Depot}: 0.5
    - Picking time t_{Pick}: 15 (seconds)
    - Variability Picking time c^2_{Pick}: 0.25, 0.5, 0.75

Overall Accuracy

In total, *experiment set 2* includes 3240 experiments and the histogram of deviations between the results of our new integrated approach and simulation is given in figure 6.3. The mean error $\bar{\varepsilon}$ is 2.2% and it is obvious that the new approach does not systematically under- or overestimate the throughput. The few experiments with an underestimation of $\varepsilon <$ -10% stem from experiments with $c^2_{Pick} = 0.25$. There is almost no overestimation with $\varepsilon > 10\%$ and deviations seem to be rather normally

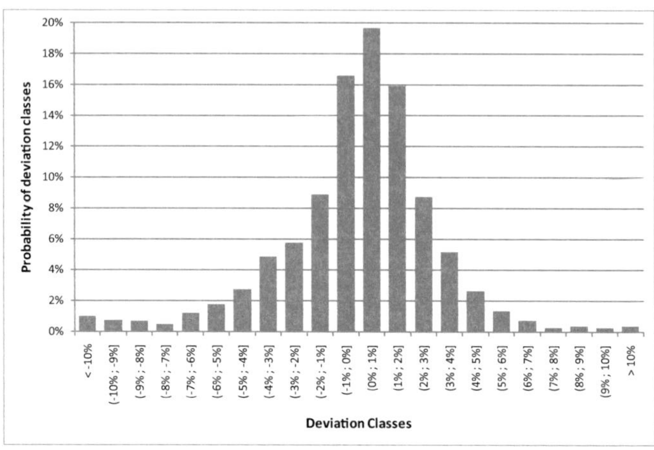

Figure 6.3: Deviation between new approach and simulation - *experiment set 2*

distributed around the value of 0%. The important error ranges and quartiles are given by:

- $P(\varepsilon \leq |5\%|) = 91\%$
- $P(\varepsilon \leq |10\%|) = 99\%$
- $Min = -12.7\%$; $Q_1 = -1.3\%$; $Q_2 = 0.3\%$; $Q_3 = 1.6\%$; $Max = 13.5\%$

Overall, we get a very good approximation accuracy for *experiment set 2*. Again, we classify the experiments by certain parameters to get a better idea on how deviations develop for certain parameter combinations and which parameter settings might produce a runaway value.

Accuracy Based on Storage Location Assignment Policy

For this parameter we can identify trends similar to those already observed for the *(extended) experiment set 1*. The approximation works very accurate for random storage. The results for medium-skewed class-based storage are also very good as the majority of experiments performs within a |5%|-error range. For high-skewed class-bases storage, the approach has a slight tendency to result in an underestimation, even though for most cases, the accuracy is within a |10%|-error range. Table 6.9 summarizes the key accuracy data.

SLAP	$P(\varepsilon \leq |5\%|)$	$P(\varepsilon \leq |10\%|)$	Min	Q_1	Q_2	Q_3	Max
Random	100	100	-3.4	-0.3	0.5	1.3	3.8
Medium-Skewed	97	100	-7.4	-1	0.7	2.2	6.6
High-Skewed	75	96	-12.7	-3.7	-1	1.7	13.5
	all values in %						

Table 6.9: Accuracy characteristics of the new approach for different storage policies - *experiment set 2*

Accuracy Based on Picker Density

As for the other experiment sets, we can observe the best accuracy for small and medium ω_{seg}. Thus the best accuracy is calculated for $\omega_{seg} = 0.06$ and we get a slightly larger deviations for $\omega_{seg} = 0.3$. Table 6.10 shows the key accuracy data for different ω_{seg}.

ω_{seg}	$P(\varepsilon \leq |5\%|)$	$P(\varepsilon \leq |10\%|)$	Min	Q_1	Q_2	Q_3	Max
0.06	99	100	-2.3	-0.3	0.7	1.6	5.3
0.12	95	100	-5.5	-0.7	0.9	2.4	6.8
0.18	88	100	-10	-1.5	0.7	2.1	8.8
0.24	87	99	-12.3	-1.7	0	1.4	10.2
0.3	86	95	-12.7	-3	-0.9	0.3	13.5
	all values in %						

Table 6.10: Accuracy characteristics of the new approach for different picker densities - *experiment set 2*

Accuracy Based on Remaining Parameters

- Number of aisles ν: there is no obvious trend to what extent the accuracy is dependent on the number of aisles ν. Thus the new integrated approach produces accurate results for all different classes of ν.

- Warehouse Shape Factor WSF: a distinct trend could not be identified and we thus subsume that accuracy is equally good for both shape factor values.

- Pick Density: the accuracy of the new approach increases for decreasing pick density. Thus we get very good results for a small pick density (0.075) with virtually no runaway values. At the same time, we recorded some for larger pick densities (0.25). Still, for this pick density, approximately 88% of experiments are within a $|5\%|$-error range.

- Representative network variability $c^2_{R,ZB}$: for $c^2_{R,ZB} \leq 1.5$, we obtain a good accuracy even though the runaway values are within this class and approximately 81% of experiments are within a $|5\%|$-error range. Results are very accurate for $c^2_{R,ZB} > 1.5$ as approximately 95% of experiments are within a $|5\%|$-error range.

- Service time $t_{s,Depot}$: as before, the service time at the depot does not seem to have an influence on accuracy.

- Picking time variability c^2_{Pick}: accuracy is best for $c^2_{Pick} = 0.5$ and we get a few errors larger than $|10\%|$ for $c^2_{Pick} = 0.25$. As before, the approach on average will result in a slight underestimation of throughput for $c^2_{Pick} = 0.25$ and will result in a slight overestimation of throughput for $c^2_{Pick} = 0.75$. The errors appear to be normally distributed around the $(0\%; 1\%]$-interval for $c^2_{Pick} = 0.5$.

The characteristic accuracy values for the remaining parameters are given in table 6.11.

Parameters	$P(\varepsilon \leq |5\%|)$	$P(\varepsilon \leq |10\%|)$
Number of aisles ν		
6	87.9	99.3
8	93.5	98.9
10	93.1	98.6
12	91.8	98.5
14	90.6	98.7
16	88.5	98
Warehouse Shape Factor WSF		
0.5	91.4	98.7
0.75	90.4	98.6
Pick Density		
0.075	91.4	99
0.15	93.3	99.8
0.25	88	97.1
Representative network variability $c^2_{R,ZB}$		
≤ 1.5	80.7	96.4
> 1.5	94.8	99.5
Service Time $t_{s,Depot}$		
1	91	98.6
15	90.8	98.7
Picking Time variability c^2_{Pick}		
0.25	87.4	97
0.5	94	100
0.75	91.2	99.1
	all values in %	

Table 6.11: Accuracy characteristics of the new approach for several input parameters - *experiment set 2*

6.1.4 Accuracy Comparison: New Approach vs. Rall's Method

In order to find out to what extent the presented approach improves the accuracy, we directly compare our new integrated approach to Rall's method. In this chapter, we will measure the error as the absolute value of the percentage deviation:

$$|\varepsilon| = 100 \cdot \left| \frac{\lambda_{\text{analytic}} - \lambda_{\text{simulative}}}{\lambda_{\text{simulative}}} \right|$$

Experiment Set 1 and Extended Experiment Set 1

We first analyzed 14580 experiments of total experiment set 1 (original and extension) and raised the following two questions for each experiment:

- Does the new approach result in a better accuracy, i.e. $|\varepsilon_{NewApproach}| < |\varepsilon_{Rall}|$?

- If improving the accuracy, how big is the improvement, i.e. how big is the difference between the two errors $|\varepsilon_{Rall}| - |\varepsilon_{NewApproach}|$?

In total, the new approach performed better in 81.7% of experiments while Rall's approach was more accurate in 18.3% of experiments.

We add some additional information to these numbers. First, we classify all experiments in two groups, namely "New Approach better" and "Rall better" and analyze the accuracy of the new integrated approach separately for these two groups. We find out that the accuracy is very good even for those experiments, where Rall's method resulted in a better accuracy:

- $P(|\varepsilon_{NewApproach}| \leq 5\%|\text{New Approach better}) = 90.1\%$
- $P(|\varepsilon_{NewApproach}| \leq 10\%|\text{New Approach better}) = 98.3\%$
- $P(|\varepsilon_{NewApproach}| \leq 5\%|\text{Rall better}) = 91.6\%$
- $P(|\varepsilon_{NewApproach}| \leq 10\%|\text{Rall better}) = 99.5\%$

The second question is concerned with the size of the improvement. Again, we classify the experiments in two groups and analyzed for each experiment to what extent an accuracy improvement could be obtained. Figure 6.4 shows the histogram of accuracy improvements.

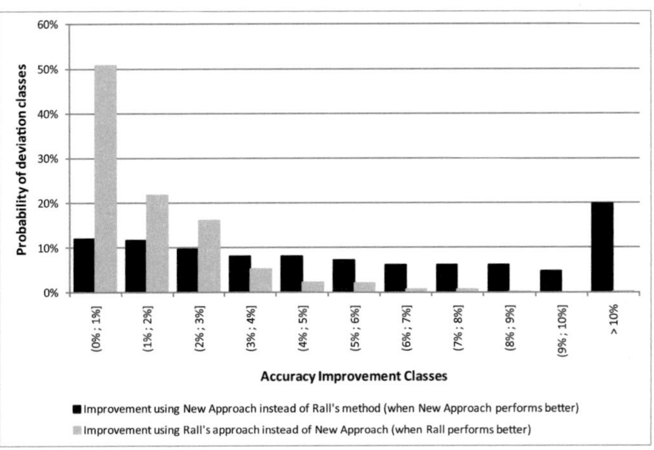

Figure 6.4: Improvement of approximation accuracy when using different analytical approaches

For experiments where the new approach resulted in a better accuracy (black bars), the improvement (by using the new integrated approach instead of Rall) is of different sizes and larger than 10% for every fifth experiment. In contrast, for experiments where Rall's method resulted in a better accuracy (grey bars), the improvement (by using Rall instead of the new integrated approach) is typically rather small with the majority of experiments resulting in an improvement of less than 1%. Even with Rall's method performing better for some experiments, the accuracy difference between both approaches is rather small and the new approach also performs very well. Thus we do not have to use different approaches for different input parameters because the new integrated approach performs well for the whole range of experiments.

Finally, we give some more information on the characteristics of accuracy improvement:

- Average accuracy improvement when using new approach instead of Rall's method for experiments where new approach performs better: 8.9%

- Average accuracy improvement when using Rall's method instead of new approach for experiments where Rall's method performs better: 1.5%

- Percentage of experiments with an accuracy improvement of more than 5% when using new approach: 50.4%

- Percentage of experiments with an accuracy improvement of more than 5% when using Rall's method: 3.6%

Experiment Set 2

We also analyzed the 3240 experiments of set 2. The new integrated approach resulted in a better accuracy for 84.8% of experiments while Rall's method was better in 15.2% of experiments.

On average, the accuracy improvement of the new approach is 7.4% and almost 50% of experiments were improved by more than 5%. For those experiments with Rall performing better, the average improvement was 2.3% and 13% of experiments had an improvement larger than 5%.

6.1.5 Validation Summary

We briefly outline the key findings of our validation procedure:

- The overall accuracy of the new integrated approach is very good, as 90% of all experiments analyzed had an error $\varepsilon \leq |5\%|$ and 99% had an error $\varepsilon \leq |10\%|$.

- The approach has an excellent accuracy for random storage policy as all respective experiments had an error $\varepsilon \leq |5\%|$.

- For class-based storage policy, two parameters have a crucial influence on accuracy: the Gini coefficient G^*, characterizing the skewness of the distribution of pick frequencies across the segments and aisles as well as the picker density ω_{seg}, characterizing the ratio of pickers to segments in the system. For larger G^*, accuracy in general does not significantly decrease. However, we experience a few runaway values, with throughput being either underestimated for small picking time variabilities ($c_{Pick}^2 = 0.25$) or overestimated for larger picking time variabilities ($c_{Pick}^2 = 0.75$). The same holds true for larger ω_{seg}.

- The remaining system-characterizing parameters do not have a crucial influence on the accuracy.

- The few runaway values can never be solely traced back to one particular parameter but result from a distinct combination of different parameters, e.g. a setting with high-skewed class-based storage, large ω_{seg} and small pick densities.

- The approach works for a wide spectrum of system-characterizing parameters. In particular, the approach also calculates very good results for input parameters that were not specifically used in the process of deriving the correction factors f_{ZB} and f_{Marie}.

6.2 Application

The objective of this chapter is to assess the magnitude of effects that congestion has on throughput in manual order picking systems. We will mainly discuss the percentage of throughput that is lost due to congestion, as calculated by:

$$\delta_{TP} = \frac{\lambda_{Linear} - \lambda_{OPS}}{\lambda_{Linear}}$$

where λ_{Linear} is the throughput of a single-picker system multiplied with the number of pickers and the number of picks, thus resulting in a linear increase of throughput. The lost throughput is of great interest to any system planer as it concerns two key questions:

- For a certain number of pickers and despite congestion, will throughput still be large enough to meet customer requirements or do we (at least temporarily) need to add additional workers to the system?

- How much money is spent on pickers waiting for each other and how does this number compare to the savings on surface costs which are ascribed to narrow aisles?

In the following, we analyze the effects of congestion in various systems by reusing the experiment sets already applied in the validation chapter. In total, we evaluated 17820 experiments, covering a wide spectrum of potential input parameter configurations. We first present some general observations about throughput losses followed by an arbitrary example which enables us to illustrate some of our typical findings with the use of graphs. Afterwards, we analyze the effects subject to different parameters.

6.2.1 General Characteristics of Throughput Loss

Calculating the average throughput loss for all experiments results in a value of 30.81%, thus on average almost a third of throughput is lost in narrow-aisle systems. Figure 6.5 illustrates the distribution of deviation classes in intervals of 5%. Additionally, the black line represents the cumulative probability of throughput losses. These occur in several classes, e.g. there are as many experiments with a loss of approximately 10% as there are losses of approximately 45%. We also notice that more than 80% of experiments have a throughput loss of at least 15%.

6.2.2 Exemplary Throughput Trends for Random Storage

To illustrate typical effects for different storage policies, we first choose an arbitrary system and show throughput trends in a graph. The chosen system has $\nu = 10$ aisles and a warehouse shape factor $WSF = \frac{2}{3}$, resulting in $h = 10$ segments per aisle[2]. Figure 6.6 shows the throughput trends for random storage.

---

[2]Other parameters include: SLAP = Random and high-skewed class-based storage, Pick Density = $\frac{1}{20}$, $t_{s,Depot} = 5s$, $c^2_{Depot} = 0.5$, $t_{Pick} = 15s$, $c^2_{Pick} = 0.5$ and picker density $\omega_{seg} = \frac{1}{100}, \frac{2}{100}, \frac{3}{100}, \frac{4}{100}, \cdots \frac{33}{100}$, i.e. we will analyze $1, 2, 3, 4, ..., 33$ pickers

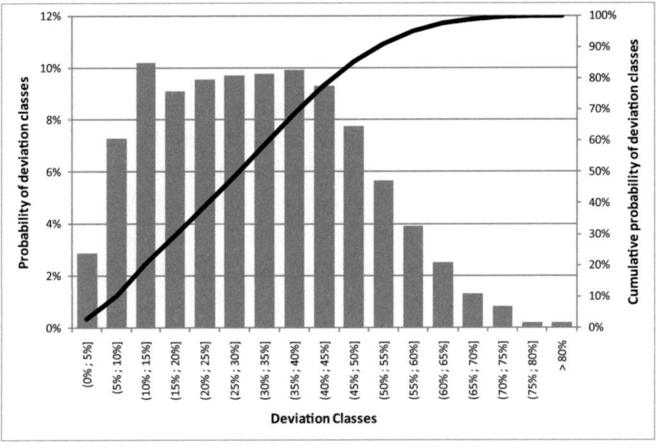

Figure 6.5: Deviation of throughput losses

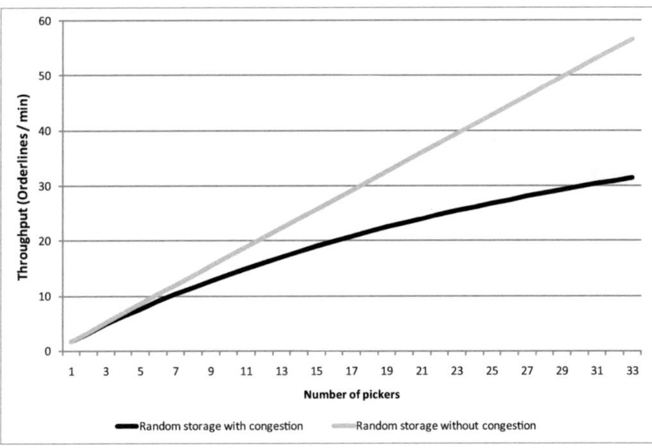

Figure 6.6: Throughput trends with/without congestion consideration - random storage

In systems without congestion consideration it is a straight line while it is increasing on a diminishing scale if pickers block each other. Losses are non-negligible even for a small number of pickers. For $K = 2$ pickers there is a throughput loss δ_{TP} of approximately 4% and for $K = 5$ pickers we already experience losses δ_{TP} of approximately 10%. This means that each picker has to spend some 10% of his total order picking time waiting for other pickers, i.e. the picker is performing tasks like picking and walking in 90% of the time. Note that for the maximum number of pickers in this example, $K = 33$, throughput losses δ_{TP} will be almost 45%. For random storage we also observe that each extra picker adds some throughput, i.e. the additional throughput generated by one extra picker compensates the additional losses which stem from an increase of congestion due to the presence of that extra picker.

6.2.3 Exemplary Throughput Trends for within-aisle Class-Based Storage

The throughput trend for the system using within-aisle class-based storage is shown in figure 6.7.

For $K = 2$ pickers the throughput loss δ_{TP} will be almost 6%. For $K = 5$ pickers, the loss δ_{TP} will already be approximately 16% and for $K = 33$, the loss δ_{TP} will be 75%. Furthermore, an additional picker will not necessarily result in a throughput gain. In a range of $K = 21, 22, ..., 25$ throughput will remain on a certain level of saturation. It will start to decrease when further increasing the number of pickers. The additional throughput of any extra picker will then be overcompensated by increased congestion due to the additional blocking situations.

6.2.4 Exemplary Comparison of Storage Policies

The comparison of different storage location assignment policies is an issue worth some more discussion. For order picking processes, it is commonly assumed that class-based storage will result in better throughput compared to random storage, because the picker presumably has to visit fewer aisles and therefore travel less distance. Consequently, the process

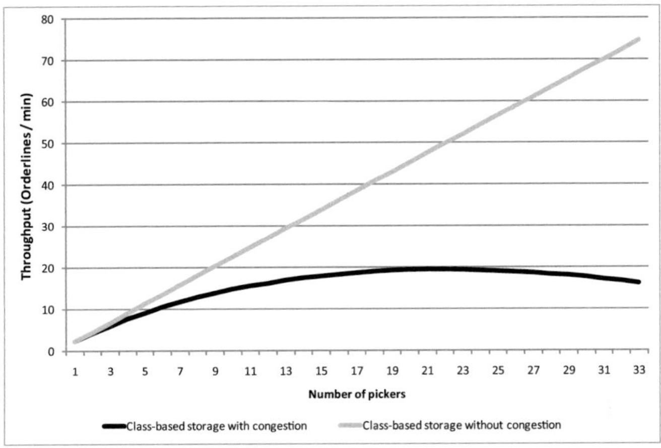

Figure 6.7: Throughput trends with/without congestion consideration - class-based storage

of allocating high runners close to the depot is surely one of the first improvement projects carried out in many order picking operations. For systems with narrow aisles, we partly question this strategy. As we can see in figure 6.8, there is a certain picker density level ω_{seg}^{SLAP}, which indicates a change of the preferable storage policy.

For $K \leq 12$, a class-based storage policy results in a better performance. In contrast, random storage will yield a higher throughput for $K \geq 13$. Thus for some picker density scenarios it is obviously better to have pickers travel larger distances instead of having them block each other in the aisles closest to the depot.

However, we need to interpret these results carefully as we are comparing two storage policies with different underlying assumptions. Remember that for random storage we assumed that each segment has the same probability of being the location of one pick. For high-skewed class-based storage we assumed that 80% of picks were located at 20% of the segments in aisles closest to the depot. We have to realize that the pick

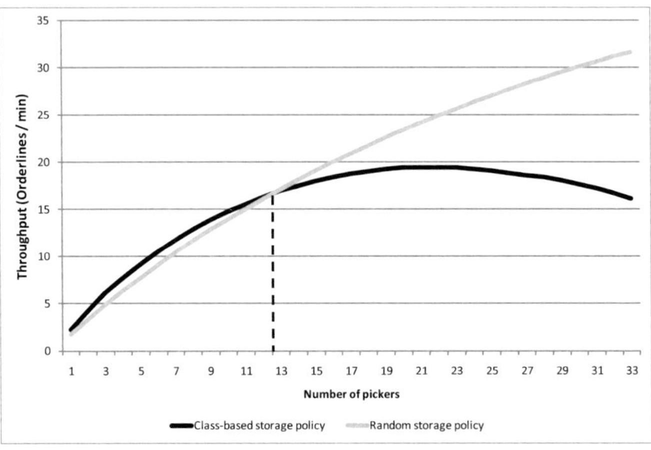

Figure 6.8: Comparison of throughput trends for different storage policies

probabilities of individual items are determined by the customer, *not* the operator of the order picking system. In case we have a given skewness in items' demand, the following two options are feasbile:

- Option 1: locate the items with high demand in the aisles closest to the depot (as we have assumed in class-based storage policy).

- Option 2: randomly select some locations to distribute high-runners evenly over the whole system. Despite the randomness of selecting these locations, this strategy is *not* the same as the random storage policy in the classical sense. This is because some segments would feature higher pick probabilities than others which is contrary to our assumption of equal pick probabilities for all segments.

A system that uses the second option will have different visit ratios than those calculated in chapters 4.5 and 4.6 and we can therefore not directly apply our models. However they can still be used as an approximation to estimate the throughput. The value of ω_{seg}^{SLAP} resulting from our models gives us an idea to what extent option 2 might be preferable to option 1. If ω_{seg}^{SLAP} is rather small, it is assumed that a more even distribution of high-runners would have a positive effect on throughput as this would ease heavy traffic in the aisles closest to the depot.

6.2.5 General Effects of Storage Policy

Expectedly, the storage location assignment policy has quite an influence on throughput losses $\bar{\delta}_{TP}$. There will be increased congestion with more picks being located close to the depot. On average, there will be a 27%-loss for random storage, a 30%-loss for medium-skewed within-aisle class-based storage and we experience an average loss of 36% for high-skewed within-aisle class-based storage. Table 6.12 also illustrates the characteristic quartiles of the losses' distributions.

SLAP	$\bar{\delta}_{TP}$	Min	Q_1	Q_2	Q_3	Max
Random	27.0	0	15.0	27.2	38.1	63.3
Medium-skewed	29.8	0	15.5	29.1	42.2	70.3
High-skewed	36.1	0	20.5	36.3	50.6	84.2
	all values in %					

Table 6.12: Throughput losses $\bar{\delta}_{TP}$ subject to different storage policies

We will also briefly focus on the critical level ω_{seg}^{SLAP}, which indicates a change of the preferable storage policy. When comparing random storage to a medium-skewed class-based storage, the average level will be $\bar{\omega}_{seg}^{SLAP} \approx 24.5\%$. For high-skewed class-based storage the average critical level will come somewhat earlier at $\bar{\omega}_{seg}^{SLAP} \approx 21.3\%$.

The pick density also has an influence on the critical level ω_{seg}^{SLAP}. It tends to be smaller for smaller number of picks and larger for larger number of picks, i.e. the critical number of pickers is reached quicker for fewer picks. This seems logical because for random storage a large number of picks will likely result in large travel times. When involving large travel times, it will be more difficult for a random storage policy to gain the advantage over within-aisle class-based storage.

6.2.6 General Effects of Order Batching

The literature states that order batching reduces travel time in single-picker operations (see chapter 2.3.3). We assume that batching involves a simple collection of orders on a first-come-first-serve basis. As an example we compare a system with a certain pick density to the same system with another pick density, thus assuming the collection of several orders that are picked simultaneously. The exemplary system configuration is reused to illustrate our findings. Figure 6.9 shows the throughput trends of the exemplary system for different number of picks, which is essentially the same as comparing non-batching (1 order with 5 picks) to batching (4 orders with 5 picks each).

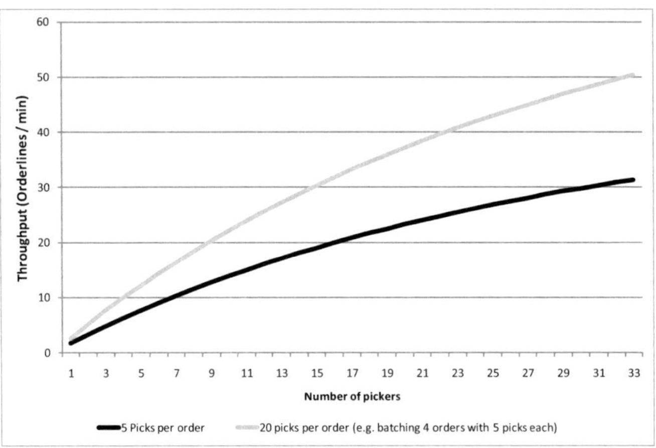

Figure 6.9: Comparison of throughput trends for order batching

In general, order batching has a positive effect on throughput in narrow-aisle systems. For each parameter constellation, throughput could be increased when picking more order lines per tour. For almost 90% of experiments, order batching also led to a reduced percentage of time a picker is blocked. Thus, the additional stops in front of racks mostly have a calming effect on the overall flow of pickers. We illustrate this by considering $\lambda_{OPS} \approx 7.6$ in figure 6.9 which is reached with $K = 5$ for non-batching and $K = 3$ for batching. Note that the percentage of time lost is 10.6% for non-batching, while being only 5.3% for batching. For most experiments, the increase in throughput stems not only from travel-time savings (based on reduced distance per order line) but can also be traced back to the reduced time a picker is blocked. For some 10% of experiments though, the percentage of time pickers are blocked is higher if batching is in use. Still, the savings on travel time compensate these increased losses to make batching favourable as overall throughput can be increased.

The average throughput losses are given in table 6.13[3]. Also included in this table are the results of two further analysis. First, we seek to find out the average percentage of throughput we could gain by applying a batching rule. For this purpose, we directly compare two experiments, which only differ in the value of pick density. For example, we assess the throughput for a certain set of input parameters with a pick density of 0.1 and compare it to the throughput of the benchmark pick density of 0.2. Secondly, we analyze how much less *percentage of time blocked* we would experience if we used a larger pick density. For a pick density of 0.1 the table reads as follows. The average throughput loss will be 32.9%. Had we used a pick density of 0.2 instead - while leaving all other parameters unchanged - we could improve throughput by 23.8% on average. Simultaneously, we would be able to reduce the percentage of time pickers are blocked by 4.9%.

Pick Density	δ_{TP}	vs. Benchmark	Throughput gain	Reduced % of time blocked
0.05	35.6%	0.2	42.1%	8.4%
0.1	32.9%	0.2	23.8%	4.9%
0.2	25.2%	-	-	-
0.075	35.0%	0.25	37.6%	8.2%
0.15	30.6%	0.25	18.2%	3.5%
0.25	27.4%	-	-	-

Table 6.13: Throughput characteristics subject to different pick densities

---

[3]The quartiles of the distribution of throughput losses subject to pick density are given in table 6.18

Finally, we should add that these findings are based on the most simple batching rule, i.e. first-come-first serve collecting. If some other batch building rule was applied, these results might change significantly. For example, a batching based on the proximity of pick locations will most likely lead to increased congestion at those locations.

6.2.7 General Effects of Picker Density

The number of pickers working in the system strongly influences the degree of congestion. As these effects are specifically distinct for different storage policies, we filter our experiments by picker density *and* storage policy. Figure 6.10 illustrates the average throughput loss $\bar{\delta}_{TP}$ when classified by these two parameters[4].

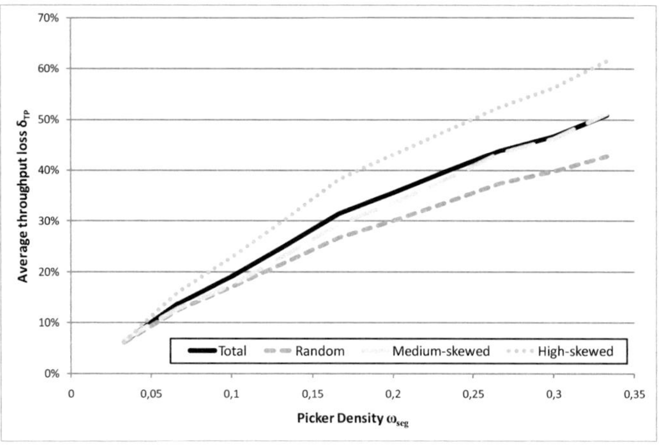

Figure 6.10: Throughput loss subject to different picker density level and storage policies

---

[4]The quartiles of the distribution of throughput losses subject to picker density are given in table 6.18.

As we would expect, the average losses are lowest for random storage and highest for high-skewed class-based storage with maximum values as high as 60%. It it noteworthy that even for very small picker densities, throughput losses quickly reach levels around 10% and should therefore not be disregarded.

We have described above (see chapter 6.2.3) a case in which an additional picker will not automatically result in an increased throughput. Analyzing this effect for all of our experiments reveals that such throughput trends are not likely to develop. For the overwhelming majority of experiments (97%) each additional picker results in additional throughput. We never experienced any decreasing throughput behavior for random storage and only very rarely for medium-skewed class-based storage. For picker density levels $\omega_{seg} \leq 0.2$ we never experienced decreasing throughputs. It does happen for some specific input parameter configurations, e.g. $\omega_{seg} > 0.2$ in combination with high-skewed class-based storage. The decrease is more likely to happen for small pick densities. We should emphasize that in our experiment sets the maximum value for picker density was $\omega_{seg} = \frac{1}{3}$. One would expect the effect of decreasing throughput to occur more frequently for picker density levels larger than that.

6.2.8 General Effects of $t_{s,Depot}$

First, we calculate the average throughput loss $\bar{\delta}_{TP}$ for the respective service times. Subsequently, we determine to what extent throughput could be increased if $t_{s,Depot}$ was reduced. Table 6.14[5] shows the results.

If $t_{s,Depot}$ is bigger than or equal to the service times of its close-by neighbors, the depot is a potential bottleneck because in contrast to other segments there will never be solely walking. For increasing picker densities the depot will ultimately become the bottleneck and we can observe a permanent queue in front of the depot with the throughput reaching a certain level of saturation. This effect occurs for random storage but is especially distinct for class-based storage. The results in table 6.14 stem from experiments with a picking time of $t_{Pick} = 15s$. It turns out that

---

[5]The quartiles of the distribution of throughput losses subject to the service time at the depot are given in table 6.18

$t_{s,Depot}$	$\bar{\delta}_{TP}$	vs. Benchmark	Throughput gain
1s	30.6%	-	-
5s	30.0%	1s	2%
15s	31.6%	1s	8.4%

Table 6.14: Throughput characteristics subject to different service times at the depot

the effect of $t_{s,Depot}$ on average throughput loss will be even bigger for $t_{Pick} = 5s$, which we have used as an input for some selected experiments. For this picking time, the difference between the service times of the depot and its close-by neighbors will be larger, thus facilitating the depot to be a bottleneck.

The general benefit of reducing the times at the depot is indisputable, e.g. resulting in a throughput increase of more than 8% if reduced from $15s$ to $1s$.

6.2.9 General Effects of the Warehouse Shape Factor

The warehouse shape factor, i.e. the ratio of the order picking system's length and width, has quite an influence on the average throughput loss. Table 6.15[6] includes the data for different shape factors.

WSF	$\frac{1}{3}$	$\frac{1}{2}$	$\frac{2}{3}$	$\frac{3}{4}$	1
$\bar{\delta}_{TP}$	22.4%	29.1%	33.3%	32.9%	36.3%

Table 6.15: Throughput characteristics subject to different warehouse shape factors

There is an obvious trend which states that on average more throughput is lost for larger shape factors. Remember that a smaller shape factor in-

---

[6]The quartiles of the distribution of throughput losses subject to the warehouse shape factor are given in table 6.18

dicates that a system tends to have many short aisles while a larger shape factor states that it tends to have few long aisles. In an order picking system with short aisles, pickers tend to have more shortcut opportunities and thus can in some cases skip heavy-traffic aisles, avoiding blocking situations. In contrast, long aisles will more likely lead to a picker being "trapped" within that aisle, resulting in more level-2-3-...-blocking situations.

To support this statement, we compare the throughput losses of scenarios that have roughly the same number of segments (νh) but are arranged with different warehouse shape factors. For all comparisons small shape factors performed better. Table 6.16 illustrates some selected examples. Note that the observed trends occur for different storage policies, being especially distinct for high-skewed class-based storage.

Segments	ν	h	WSF	$\bar{\delta}_{TP}$
16	8	2	$\frac{1}{3}$	19.0%
18	6	3	$\frac{1}{2}$	23.1%
100	10	10	$\frac{2}{3}$	31.3%
104	8	13	1	35.3%
336	16	21	$\frac{3}{4}$	35.1%
350	14	25	1	38.1%

Table 6.16: Comparison of different warehouse shape factors for approximately equal number of segments

6.2.10 General Effects of c^2_{Pick}

Concerning the variability of the picking time, we found that throughput losses tend to increase slightly with increasing c^2_{Pick}. Larger c^2_{Pick} will ultimately result in larger variabilities c^2_{WA} within the aisles. Larger variabilities obviously sometimes will force pickers to stay longer at one segment, thus increasing the likelihood of a blocking situation. Note that this only focuses on throughput *losses*. We also analyze to what extent throughput will be won/lost if we decrease/increase c^2_{Pick}. Table 6.17[7] summarizes the results. By reducing c^2_{Pick} from 0.5 to 0.25, an average

c^2_{Pick}	$\bar{\delta}_{TP}$	vs. Benchmark	Average throughput win/loss
0.25	28.8%	0.5	5.6%
0.5	30.9%	-	-
0.75	32.8%	0.5	-4.2%

Table 6.17: Throughput characteristics subject to different picking time variabilities

throughput win of almost 6% could be realized while on the other hand average throughput would be reduced by 4% if variability increased to 0.75.

The storage policy has a certain influence on these numbers. If we consider only experiments with random storage, then throughput could be increased by 3.7% for $c^2_{Pick} = 0.25$ and throughput would decrease by 3.2% for $c^2_{Pick} = 0.75$. For high-skewed class-based storage the influence is biggest. Reducing variability would result in an 8.5% increase of throughput. Contrary, we would loose 5.5% of throughput if variability increased.

---

[7]The quartiles of the distribution of throughput losses subject to picking time variability are given in table 6.18

6.2.11 Approximation Accuracy for Throughput Losses

Finally, we briefly analyze the accuracy of the new integrated approach if we concentrate on throughput *losses*. We compare δ_{TP} calculated with the new integrated approach to δ_{TP} calculated using simulation. The maximum runaway values are significantly reduced, e.g. the largest underestimation of almost 32% (throughput) is reduced to an underestimation of 8.3% (throughput loss). The mean error for all experiments of sets 1 and 2 is reduced from 2.5% (throughput) to 1.6% (throughput loss). Overall the approximation accuracy for the percentage of time pickers are blocked is even better than the results in the validation chapter (6.1) suggest.

Figure 6.11 shows the probabilities of an arbitrary experiment to be in a certain deviation class. We see that for almost 50% of experiments the throughput loss δ_{TP} is approximated with an error smaller than or equal to 1%. In total, 97% of experiments have an error smaller than or equal to 5%.

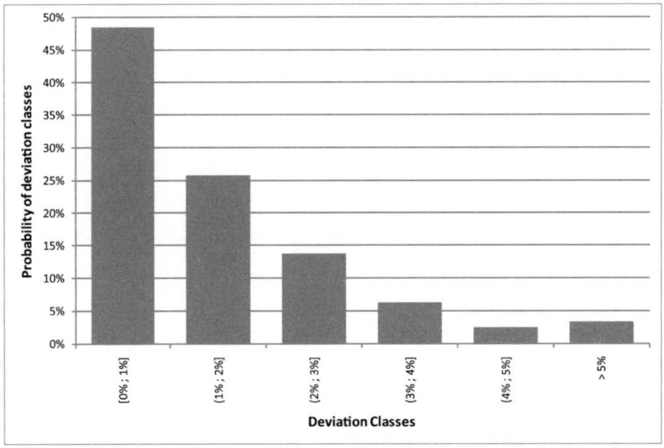

Figure 6.11: Accuracy of the approximation of throughput losses

Parameter	$\bar{\delta}_{TP}$	Min	Q_1	Q_2	Q_3	Max
Pick Density (Order Batching)						
0.05	35.6%	0%	20.4%	35.1%	49.9%	84.2%
0.075	35.0%	5.9%	21.9%	35.0%	46.7%	72.2%
0.1	32.9%	0%	19.0%	32.8%	45.5%	75.4%
0.15	30.6%	5.8%	19.5%	30.4%	41.0%	63.0%
0.2	25.2%	0%	12.3%	24.0%	37.5%	64.2%
0.25	27.4%	4.9%	17.1%	27.0%	36.6%	57.0%
Picker Density ω_{seg}						
0.0333	6.3%	0%	3.9%	7.3%	9.2%	14.8%
0.06	13.4%	6.2%	11.2%	13.1%	15.4%	22.8%
0.0667	13.8%	0%	11.2%	14.0%	17.2%	27.6%
0.1	19.1%	0%	15.0%	19.8%	24.0%	37.9%
0.12	22.5%	4.9%	18.7%	22.7%	26.6%	39.1%
0.1333	25.2%	0%	20.9%	25.9%	30.8%	46.3%
0.1667	31.5%	4.5%	25.8%	31.6%	37.2%	54.0%
0.18	31.9%	9.9%	26.9%	31.8%	36.8%	51.5%
0.2	35.5%	4.5%	29.5%	35.6%	41.9%	60.0%
0.2333	39.7%	4.4%	33.4%	40.6%	47.3%	65.7%
0.24	39.8%	15.0%	33.8%	39.8%	45.9%	62.1%
0.2667	43.8%	4.5%	37.6%	44.7%	52.1%	71.7%
0.3	46.7%	4.4%	39.7%	47.1%	55.5%	77.5%
0.3333	50.9%	4.4%	43.5%	51.8%	61.7%	84.2%
Service Time Depot $t_{s,Depot}$						
1	30.6%	0%	17.0%	30.0%	42.5%	84.2%
5	30.0%	0%	15.8%	29.5%	42.6%	84.1%
15	31.6%	0%	17.2%	31.1%	44.3%	83.9%
Warehouse Shape Factor						
0.333	22.4%	0%	10.0%	20.7%	33.9%	70.8%
0.5	29.1%	4.9%	17.5%	28.4%	38.8%	70.7%
0.667	33.3%	0%	19.7%	33.3%	45.4%	82.4%
0.75	32.9%	6.9%	20.7%	32.7%	43.5%	72.2%
1	36.3%	2.8%	22.4%	36.5%	48.9%	84.2%
Picking Time Variability c_{Pick}^2						
0.25	28.8%	0%	15.0%	28.1%	40.7%	82.4%
0.5	30.9%	0%	17.0%	30.5%	43.2%	83.4%
0.75	32.8%	0%	18.9%	32.9%	45.5%	84.2%

Table 6.18: Average values and quartiles of throughput losses for different input parameters

6.2.12 Key Observations on Congestion Effects

We summarize the most important effects of congestion in manual order picking systems with narrow aisles. Planers and operators should bear these observations in mind in order to get a realistic estimation of the expected performance. Obviously, these findings are based on the results we calculated using our model which itself uses some assumptions on routing and storage policy. If those change, for example using return routing policy and across-aisle class-based storage, the following observations might partly change.

- Congestion always prevents throughput from increasing linearly in the number of pickers. Instead the throughput trend will be increasing on a diminishing scale. Even for very small picker densities ω_{seg}, the effects are noticeable. For very large levels of ω_{seg}, throughput might even be decreasing if additional pickers are put into the system. In any case, picker productivity (measured in picked order lines per man hour) decreases with every additional picker.

- On average, congestion will account for a throughput loss of roughly 30%. Some parameter scenarios yield very high throughput losses but still an 80%-majority of experiments have a loss of at least 15%.

- Throughput losses will be fewest for random storage policy because blocking situations will be less likely and mostly involve no more than two pickers. Losses are bigger for within-aisle class-based storage policies because many pickers typically sojourn only certain parts of the whole area, resulting in heavy-traffic aisles.

- Up to a critical picker density level ω_{seg}^{SLAP}, a within-aisle class-based storage policy will result in higher throughputs compared to a random policy. Beyond the critical level however, throughput of within-aisle class-based storage systems is actually lower than that of a random storage. This is an important finding as existing literature not considering congestion mostly reports on the unconditional advantages of class-based storage compared to random storage. From a large set of experiments, we derived the average critical level for high-skewed class-based storage as $\bar{\omega}_{seg}^{SLAP} \approx 0.21$. From this number we can indirectly derive the guideline that if the ratio of pickers to picking segments is larger than $1:5$, an even distribution of frequently demanded items across the whole system is likely to perform better than clustering them in one or several high-runner-aisles.

- Concerning order batching based on first-come-first-serve collection of orders, we support the statements of existing literature, claiming its positive effects. In narrow-aisle systems, the savings on travel-time also justify this strategy. Furthermore in most cases the increased picking activities also have a calming effect on overall picker flow and pickers spend less of their time waiting for other pickers. Of course batching is generally only recommended if a required subsequent sorting process is efficient enough and if resulting order sojourn times are acceptable. We note that other batch building rules (e.g. proximity of orders) which were not part of the analysis might have a negative influence on throughput.

- The administrative time at the depot can have an influence on the systems' performance. Whenever the time at the depot is comparatively much larger than the time needed at its close-by neighbors, the depot has an immediate potential to be the bottleneck of the whole system. This trivial statement is especially crucial for the depot because all pickers use it to start and end their tours. Investments to cut short administrative times still have to be mindfully considered because its effects might be overestimated. In our experiments, the 93%-reduction from 15s to 1s resulted in an average throughput gain of approximately 8%.

- The average picking times surely have a bigger impact than the time at the depot. In some of our experiments, reducing it by 67% from $15s$ to $5s$ resulted in a throughput increase of almost 80%. Needless to say this effect will strongly depend on the number of picks per order and the ratio of picking time to order sojourn time.

- The variability of picking time also has an influence on throughput. However, throughput gains or losses are not as significant as those based on picking time. By cutting the variability from 0.5 to 0.25, throughput losses can be slightly increased and the average throughput win is almost 6%. The positive impact of reducing variabilities is higher for class-based storage, resulting in 8% additional throughput. Still, it seems to take quite an effort to cut variability by 50% and "only" produce a throughput gain of 6 or 8% respectively. One should be aware tough that variability of the picking time ultimately influences the variability of total order sojourn time. If we look beyond the borders of the order picking system, we might find that a reduced variability of its output will help other succeeding processes perform better because material actually flows more consistently through the whole warehouse.

- There is an obvious trend stating that congestion tends to be heavier in systems with larger warehouse shape factors. This means that systems with few long aisles tend to suffer more from congestion than systems with many short aisles. In particular pickers will be less likely to skip aisles and will therefore sometimes by "trapped" in aisles. This effect is especially distinct for high-skewed class-based storage. We therefore recommend to design narrow-aisle systems in a way they have more width W than length L, ideally with $W \geq 2L$. Note that this is in contrast to the guideline given by Hwang et al. (2004, p. 3883). It states that a system should be designed such that $W \approx \frac{L}{2}$, i.e. a few long aisles should be used (see aisle configuration problem in chapter 2.2.2). This represents yet another example that classic guidelines need to be considered carefully when designing and operating narrow-aisles systems.

7 Conclusion

7.1 Summary

In manual order picking systems with narrow aisles, pickers cannot pass each other. Consequently there are interdependencies between the pickers that negatively influence throughput and thus the performance of such systems. In particular, pickers are interrupted by other pickers and have to wait for each other. This results in blocking situations and congestion. For such systems, throughput will be falsely estimated if the result of a single-picker travel-time model is simply multiplied with the number of pickers. Instead, models explicitly have to consider interdependencies and resulting congestion effects.

We have presented an analytical approach to calculate throughput in manual order picking systems by means of continuous time queueing theory. The model was specifically designed to consider congestion, thus quantifying the effects of picker interferences. It is well-suited to support the rough planning phase in which different design alternatives need to be evaluated. At this point an analytical approach is preferred over simulation because it quickly calculates key performance indicators, whereas simulation consumes more time. Typically, analytical models are used in the first planning loop and some promising parameter configurations are subsequently analyzed in more detail using simulation.

The presented model allows for the consideration of several order picking input parameters, such as random or class-based storage policy, picking and administrative times as well as traversal routing with and without aisle-skipping. The choice of parameters is a major advantage compared to existing approaches, which offer rather few possibilities to analyze various configurations because they are based on tight assumptions.

We modeled an order picking system as a closed queueing network in which the systems' resources are represented by elementary queueing systems. For instance, resources include a rack column within an aisle or the depot and the corresponding space in front of it. Thus the space pickers use for walking or riding can be divided into segments, which each can hold at maximum one picker at a time. Each segment is modeled by an elementary zero-buffer queueing system with one server. The absence of a waiting room enables the modeling of blocking situations and thus congestion. We derived formulas to determine transition probabilities, i.e. routing probabilities for segments at which pickers can decide on the direction of travel. This step is pivotal for the modeling of routing and storage location assignment policies. Because picking times are assumed to be generally distributed, the model is completed by the derivation of formulas for the expected service time as well as service time variability for each segment.

We have pointed out that closed queueing networks with unlimited buffers are basically suitable to model blocking situations involving two pickers. If more than two pickers are involved, we have to use networks with limited buffers to model congestion appropriately. We present a new integrated approximation approach for the latter as no existing method is precisely fitting the requirements. The basic idea is to model the decrease of throughput which is caused by blocking situations involving more than two pickers as a reduction of the number of pickers working in the system. This reduction is achieved by applying the correction factor f_{ZB}. The reduced number of pickers is then used in the network in which the zero-buffers are replaced with unlimited buffers. Such networks can be analyzed with the well-known method of Marie. However, for some parameter configurations, the accuracy of Marie's method needs to be improved by applying the correction factor f_{Marie}. Both f_{ZB} and f_{Marie} are obtained by closed-form expressions depending on characteristic system parameters.

The procedure was validated by applying it to a large set of 17820 experiments. Throughput was calculated and compared to the results of simulation. We found that 90% of experiments had an approximation error $\varepsilon \leq |5\%|$ and 99% of experiments had an approximation error $\varepsilon \leq |10\%|$.

Thus the new approach calculates throughput values of very good accuracy. Within a rough planning phase, the observed errors are acceptable.

To gain some insights on system behavior when congestion is occurring, we calculated the percentage of throughput lost by comparing results of our new integrated approach to calculations not considering congestion. First, the new method was used for the large experiment set to calculate throughput for K pickers. Subsequently, to obtain throughput of a system without any congestion, we calculated the throughput for 1 picker and multiplied it with K. The comparison showed that throughput losses can be of significant size. The average loss is approximately 31% and an 80%-majority of experiments have a loss of at least 15%. Unsurprisingly, the loss increases with increasing number of pickers.

Concerning the storage location assignment policy, the average loss is 36% for a high-skewed class-based storage policy and 27% for random storage. In systems without congestion, it is commonly assumed that class-based storage always outperforms random storage in terms of throughput because travel distances will be smaller. Our results suggest that this guideline can only be partly applied to order picking systems with narrow aisles. Class-based storage will always perform better for a small number of pickers. In contrast, we can identify scenarios where random storage performed better once a critical number of pickers was exceeded. In such cases the additional pickers in a system with class-based storage will cause comparatively more congestion than in systems with random storage. In fact the pickers will walk less distance, yet they will spend more time waiting, which will overcompensate the combined walking and waiting time in a system with random storage. In some scenarios with class-based storage, we could also identify that an additional picker does not necessarily create additional throughput. Thus at some stage the throughput trends might also be decreasing when extra pickers are put into the system.

Order batching will have a positive effect on throughput in narrow-aisle systems as travel time per order line decreases. Moreover we also experienced a calming effect of batching for most of our scenarios as percentage throughput loss will be smaller compared to non-batching.

Another influential parameter is the warehouse shape factor, defining the ratio between length and width of the order picking system. We found that more congestion will occur in systems with larger warehouse shape factors. We compared different design alternatives and recommend narrow-aisle systems to be built with many short aisles instead of a few long aisles. This guideline is especially important for systems with class-based storage. Again this recommendation is contrary to statements in the existing literature which were derived from models not considering congestion effects.

Although our model is applicable to far more scenarios than any of the existing analytical approaches, we can still identify the following weaknesses of our approach. Queueing models are mostly used in the rough planning phase as it is unlikely that they provide the necessary level of detail for more precise analysis. For these we still have to rely on other methodologies, e.g. simulation. Even though the overall accuracy is very satisfying, we experienced a few runaway values as high as approximately $|30\%|$. These stem from the formulas of the correction factors, which were derived from empirical data. Even though we have used a wide range of parameters, we still cannot guarantee the factors' fitness for every single possible parameter configuration. The last weakness we should discuss is the fact that the results of the developed approach are average values. Consequently, we do not obtain any information about the distribution of performance indicators. Some interpretations might therefore be incomplete. For example, the average order cycle time might be satisfactory while the cycle time distribution would reveal that a certain percentage of orders has unacceptably long cycle times. By using average values for the input parameters, we might also loose some information on system behavior, e.g. if picking times have a multimodal distribution. In contrast to continuous time queueing models which have been used in this work, discrete time queueing models use distributions and therefore seem to be a promising alternative. However, we should bear in mind that the application of discrete time queueing models to large and complex networks is limited. Thus results for at least some of our experiments could not have been calculated with an acceptable effort.

7.2 Outlook

The presented approximation approach constitutes the first thorough application of queueing models to narrow-aisle order picking systems, enabling the calculation of throughput with congestion consideration. Consequently, many open research questions remain and thus there are several possible extensions to this approach. The one-way traversal routing policy was implemented by calculating transition probabilities. Using other ways to derive these values, we could possibly implement other routing strategies, e.g. return strategy. However, one would have to incorporate a solution for opposing traffic as we have pointed out that such opposed movements might require priority rules defining which picker gets to go first. The location of the depot could also be altered by modifying the transition probabilities. Loosening the assumption of a fix number of picks per order will have some influence on transition probabilities as well as service time parameters. A major advancement could be achieved by implementing the passing process of pickers within an aisle, thus not solely considering narrow aisles. This could involve having a maximum possible number of pickers that can simultaneously use a segment. For example one picker could just pass the segment if only one other picker was using it and one picker would have to wait if two others were already using it. Furthermore we could allow for each aisle to have its individual width. This way heavy-traffic aisles close to the depot could be wide enough to allow for passing while low-traffic aisles far from the depot could be narrow. Presumably some of the congestion problems in systems with class-based storage and a high number of pickers could be solved by this strategy. Concerning the modeling of this extension, multiple-server queueing systems seem to be a promising approach. Finally, an optimization model which uses the results of the new integrated approach and provides an optimal set of system parameters subject to certain throughput and layout requirements could further help planners to efficiently decide on layout and operational strategies.

Glossary of Notation

A_x — decision point of type A, immediately preceding the even aisle $2x + 2$

α — index for an aisle

α_{even} — index for an even aisle

α_{next} — index of the next aisle a picker has to enter

α_{odd} — index for an odd aisle

B_i — capacity of queueing system i, i.e. the sum of the number of servers m and the number of spaces in the buffer

β — index for an aisle

β_{odd} — index for an odd aisle

β_{even} — index for an even aisle

c^2 — variability / squared coefficient of variation

c^2_{Depot} — variability at the depot

c^2_{CA} — variability at a cross aisle segment

c^2_{Pick} — variability of the picking time (note that this is different from c^2_{WA} because the latter also includes the walking activity and the probability that a pick is done)

c^2_R — representative network variability, expressing to what extent the network is subject to the influences of nodes' variabilities

$c^2_{R,ZB}$ — adjusted representative network variability for zero-buffer networks

c^2_{WA} — variability at a segment within an aisle

D — stochastic administrative time at the depot

D_{Rack} — depth of a rack column; assumed to be $0.5m$

δ_{TP} — percentage of throughput loss caused by congestion; equals the percentage of time a picker is blocked

E — number of experiments

e_i	visit ratio at queueing system i, indicating the relative portion of network throughput that goes through system i
$e_{s,rel}$	relative portion of a single visit ratio e_s to total visit ratios for all segments with picking
$E(t_a)$	average interarrival time between two arriving customers
$E(t_s)$	average service time of a customer
$E(t_{s,CA})$	average service time at a segment of the cross aisle
$E(t_{s,Depot})$	average service time at the depot
$E(t_{s,WA})$	average service time at a segment within an aisle
ε	error between an analytically calculated value and a value resulting from simulation
f_{Marie}	correction factor that adjusts the number of customers in a way enabling us to use the method of Marie to estimate throughput in a network with unlimited buffers, thus compensating the errors of Marie's method
f_{Marie}^{true}	true correction factor as calculated from a simulation
f_{Rall}	correction factor that incorporates the effects of variability in the result of Akyildiz' algorithm
f_{ZB}	correction factor that adjusts the number of customers in a way enabling us to use a network with unlimited buffers to estimate the throughput in zero-buffer networks
f_{ZB}^{true}	true correction factor as calculated from a simulation
G^*	Gini coefficient describing the distribution of pickers to areas of the order picking system (0=even distribution over whole system; 1=concentrated distribution)
Γ_{DP}	number of different cases how a certain number of picks can be distributed to the different segments of the system in a way each case will lead to a visit of decision point DP
Γ_{DP}^{CB}	number of different cases how a certain number of picks can be distributed to the different segments of the system when a class-based storage policy is used

γ	index for a class in class-based storage policy
$\gamma_{p,picks}$	percentage of picks done in class γ
$\gamma_{p,seg}$	percentage of segments belonging to class γ
γ_{seg}	absolute number of segments belonging to class γ
$\gamma_{virtual}$	multiplying factor which is used to increase an aisle to its virtual size
h	number of segments per aisle. A segment is the space in front of two opposing rack columns
h_α	number of segments per aisle after virtually increasing the size of aisle α
K	number of customers in the queueing network or number of pickers
K^*	number of customers in the queueing network as calculated by Akyildiz
K^{**}	number of customers in the queueing network as calculated by Rall
K^+	adjusted number of customers enabling us to use a network with unlimited buffers to estimate the throughput in zero-buffer networks
K^+_{true}	true adjusted number of customers as calculated from a simulation
K^{++}	adjusted number of customers enabling us to use the method of Marie to estimate throughput in a network with unlimited buffers, thus compensating the approximation errors of Marie's method
K^{++}_{true}	true adjusted number of customers as calculated from a simulation
L	length of the order picking system (measured along the length of an aisle; $L = \frac{h}{\nu} \cdot L_{Rack} + 3[m]$)
L_{Rack}	length of a rack column; assumed to be $1m$
L^{CB}	number of different classes in class-based storage
λ_{Linear}	throughput in an order picking system calculated by multiplying the throughput of one customer with the number of customers
$\lambda_{Marie}(K)$	throughput in a network with unlimited buffers and K customers as calculated by Marie's method

$\lambda_{OPS}(K)$	throughput in an order picking system with K pickers; measured in completed order lines per period of time
$\lambda_{UB}(K)$	throughput in a network with unlimited buffers and K customers
$\lambda_{ZB}(K)$	throughput in a zero-buffer network with K customers
m	number of parallel servers of a queueing system
μ	service rate
μ_{CA}	service rate at a cross aisle
μ_D	mean value of the stochastic administrative time at the depot
μ_{Depot}	service rate at the depot
μ_P	mean value of the stochastic picking time
μ_{WA}	service rate at a segment within an aisle
N	number of servers in the queueing network
N_S	average number of customers in the system
N_W	average number of customers in the buffer
n	number of picks per order / number of order lines per order
ν	number of aisles in the order picking system
o	auxiliary variable for indicating aisle usage
ω	customer or picker density, defined as the ratio $\dfrac{K}{\sum_{i=1}^{N} B_i}$
ω_α	customer or picker density within an aisle α
ω_{seg}	customer or picker density based on segments, defined as the ratio $\dfrac{K}{\nu \cdot h}$
ω_{seg}^{SLAP}	picker density level where a change of the storage policy is preferable
$p_{pick,i}$	probability that an arbitrary order includes a pick at segment i
$p_{stop,i}$	probability that a picker has to stop at segment i in order to do a pick
P	stochastic picking time at a segment
$q_{j,i}$	transition probability, indicating the probability that a customer after finishing service at system j is transferred to system i
Q_x	quartile of a distribution. Q_1 is the lower quartile, i.e. the 25th percentile; Q_2 is the median, i.e. the 50th percentile; Q_3 is the upper quartile, i.e. the 75th percentile

r	Pearson's coefficient of correlation
R^2	coefficient of determination, indicating the fitness of a function
ρ	utilization of a queueing system
σ_D^2	standard deviation of the stochastic administrative time at the depot
σ_P^2	standard deviation of the stochastic picking time
t_{OPT}	total order picking time, i.e. the time to complete one picking tour staring and ending at the depot
t_{Pick}	picking time at a rack column (note that this is different from $E(t_{s,WA})$ because the latter also includes the walking activity and the probability that a pick is done)
t_w	waiting time in a queueing system
t_v	sojourn (or throughput) time
T	deterministic travel time
W	width of the order picking system (measured along the length of the total cross aisle; $W = \nu \cdot W_{Aisle} + 2 \cdot \nu \cdot D_{Rack}[m]$)
W_{Aisle}	width of an aisle; assumed to be $1m$
WSF	warehouse Shape Factor, indicating the ratio $\frac{L}{W}$
y	auxiliary variable for allocating picks to aisles
z	auxiliary variable for allocating picks to aisles

References

Agrawal, S. C., J. P. Buzen and A. W. Shum (1984, August). Response Time Preservation: a general technique for developing approximate algorithms for queueing networks. *ACM Sigmetrics Performance Evaluation Review 12*(3), p. 63–77.

Akyildiz, I. F. (1987). Exact Product Form Solution for Queueing Networks with Blocking. *IEEE Transactions on Computers 36*(1), p. 122–125.

Akyildiz, I. F. (1988a). General Closed Queueing Networks with Blocking. In: P.-J. Courtois and G. Latouche (eds.), *Performance '87, Proceedings of the 12th IFIP WG 7.3 International Symposium on Computer Performance Modelling, Measurement and Evaluation*, p. 283–303. North Holland.

Akyildiz, I. F. (1988b). On the Exact and Approximate Throughput Analysis of Closed Queuing Networks with Blocking. *IEEE Transacions on Software Engineering 14*(1), p. 62–70.

Akyildiz, I. F. (1989). Product Form Approximations for Queueing Networks with Multiple Servers and Blocking. *IEEE Transactions on Computers 38*(1), p. 99–114.

Akyildiz, I. F. (2009). Personal e-mail correspondence. February 2009.

Arnold, D. and K. Furmans (2009). *Materialfluss in Logistiksystemen* (6 ed.). VDI. Berlin: Springer.

Ashayeri, J. and L. Gelders (1985). Warehouse design optimization. *European Journal of Operational Research 21*(3), p. 285–294.

Augustin, R. and K.-J. Büscher (1982). Characteristics of the COX-distribution. *SIGMETRICS Perform. Eval. Rev. 12*, p. 22–32.

Backhaus, K., B. Erichson, W. Plinke and R. Weiber (2003). *Multivariate Analysemethoden : eine anwendungsorientierte Einführung* (10 ed.). Berlin: Springer.

Balsamo, S. (1993). Properties and analysis of queueing network models with finite capacities. *Lecture Notes in Computer Science 729*, p. 21–52.

Balsamo, S., V. de Nitte Personé and R. Onvural (2001). *Analysis of queueing networks with blocking*. Norwell, MA: Kluwer Academic Publishers.

Bartholdi, J. J. and D. D. Eisenstein (1996). A production line that balances itself. *Operations Research 44*(1), p. 21–34.

Bartholdi, J. J., D. D. Eisenstein and R. D. Foley (2001). Performance of bucket brigades when work is stochastic. *Operations Research 49*(5), p. 710–719.

Bartholdi, J. J. and K. R. Gue (2000). Reducing labor costs in an LTL crossdocking terminal. *Operations Research 48*(6), p. 823–832.

Bartholdi, J. J. I. and S. T. Hackman (2007). Warehouse & Distribution Science. http://www.warehouse-science.com.

Bassan, Y., Y. Roll and M. J. Rosenblatt (1980). Internal Layout Design of a Warehouse. *AIIE Transactions 12*(4), p. 317–322.

Berry, J. (1968). Elements of warehouse layout. *The International Journal of Production Research 7*(2), p. 105–121.

Bolch, G., S. Greiner, H. de Meer and K. S. Trivedi (1998). *Queueing Networks and Markov Chains: modeling and performance evaluation with computer science applications*. John Wiley & Sons, Inc.

Bouhchouch, A., Y. Frein and Y. Dallery (1996). Performance evaluation of closed tandem queueing networks with finite buffers. *Performance Evaluation 26*(2), p. 115 – 132.

Bronstein, I. N. and K. A. Semendjajew (1991). *Taschenbuch der Mathematik* (25 ed.). Stuttgart, Moscow: B.G. Teubner Verlagsgesellschaft and Verlag Nauka.

Buzen, J. (1973). Computational Algorithms for Closed Queueing Networks with Exponential Servers. *Comunications of the ACM 16*(9), p. 527–531.

Caron, F., G. Marchet and A. Perego (1998). Routing policies and COI-based storage policies in picker-to-part systems. *International Journal of Production Research 36*(3), p. 713–732.

Caron, F., G. Marchet and A. Perego (2000a). Layout design in manual picking systems: a simulation approach. *Integrated Manufacturing Systems 11*(2), p. 94–104.

Caron, F., G. Marchet and A. Perego (2000b). Optimal layout in low-level picker-to-part systems. *International Journal of Production Research 38*(1), p. 101–117.

Chew, E.-P. and L. C. Tang (1997). Order Picking Systems: Batching and Storage Assignment Strategies. *Computers & Industrial Engineering 33*(3-4), p. 817–820.

Chew, E. P. and L. C. Tang (1999). Travel time analysis for general item location assignment in a rectangular warehouse. *European Journal of Operational Research 112*, p. 582–597.

Chiang, W.-C., P. Kouvelis and T. L. Urban (2002). Incorporating Workflow Interference in Facility Layout Design: The Quartic Assignment Problem. *Management Science 48*(4), p. 584–590.

Chiang, W.-C., P. Kouvelis and T. L. Urban (2006). Single- and multi-objective facility layout with workflow interference considerations. *European Journal of Operational Research 174*, p. 1414–1426.

Choe, K. I. and G. Sharp (1991). Small Parts Order Picking: Design and Operation. http://www2.isye.gatech.edu/logisticstutorial/order/article.htm.

Cormier, G. and E. A. Gunn (1992). A review of warehouse models. *European Journal of Operational Research 58*, p. 3–13.

Dallery, Y. and X.-R. Cao (1992). Operational analysis of stochastic closed queueing networks. *Performance Evaluation 14*, p. 43–61.

Dallery, Y. and Y. Frein (1989). A Decomposition Method for the Approximate Analysis of Closed Queueing Networks with Blocking. In: H. Perros and T. Altiok (eds.), *Queueing Networks with Blocking*, p. 193–216. North Holland.

Dallery, Y. and D. D. Kouvatsos (1998). *Queueing Networks with Blocking*. Annals of Operations Research. Baltzer Science Publishers.

de Koster, R. (1994). Performance approximation of pick-to-belt orderpicking systems. *European Journal of Operational Research 72*, p. 558–573.

de Koster, R., T. Le-Duc and K. J. Roodbergen (2007). Design and control of warehouse order picking: A literature review. *European Journal of Operational Research 182*, p. 481–501.

de Koster, R., K. Roodbergen and E. van der Poort (1998). When to apply optimal or heuristic routing of orderpickers. In: F. B. et al. (eds.), *Advances in Distribution Logistics*, Number 460 in Lecture notes in economics and mathematical systems, p. 375–401. Berlin: Springer.

de Koster, R. and E. van der Poort (1998). Routing orderpickers in a warehouse: a comparison between optimal and heuristic solutions. *IIE Transactions 30*, p. 469–480.

de Koster, R., E. S. van der Poort and M. Wolters (1999). Efficient orderbatching methods in warehouses. *International Journal of Production Research 37*(7), p. 1479–1504.

De Souza E Silva, E., S. S. Lavenberg and R. R. Muntz (1986). A Clustering Approximation Technique for Queueing Network Models with a Large Number of Chains. *IEEE Transactions on Computers 35*, p. 419–430.

Dudley, N. (1963). Work-time distributions. *International Journal for Production Research 2*(2), p. 137–144.

Dukic, G. and C. Oluic (2007). Order-picking methods: improving order-picking efficiency. *International Journal of Logistics Systems and Management 3*(4), p. 451–460.

Fahrmeir, L., R. Künstler, I. Pigeot and G. Tutz (2003). *Statistik: der Weg zur Datenanalyse* (4 ed.). Berlin Heidelberg New York: Springer.

Fahrmeir, L., R. Künstler, I. Pigeot and G. Tutz (2007). *Statistik: Der Weg zur Datenanalyse* (6 ed.). Springer-Lehrbuch. Berlin, Heidelberg: Springer-Verlag Berlin Heidelberg.

Faißt, B. and C. R. Lippolt (2002). Staueffekte in Materialflusssystemen. *F+H Fördern und Heben 52*(8), p. 506–508.

Francis, R. L. (1967). On Some Problems of Rectangular Warehouse Design and Layout. *The Journal of Industrial Engineering 18*(10), p. 595–604.

Frazelle, E. (2002). *World-Class Warehousing and Material Handling*. New York: McGraw-Hill.

Frazelle, E. and S. T. Hackman (1994). The forward-reserve problem. In: T. Ciriani and R. Leachman (eds.), *Optimization in Industry 2*, p. 43–61. Wiley.

Furmans, K. (1992). *Ein Beitrag zur theoretischen Behandlung von Materialflußpuffern in Bediensystemnetzwerken*. Ph.D. thesis, Universität Karlsruhe.

Furmans, K. (2000). *Bedientheoretische Methoden als Hilfsmittel der Materialflußplanung*. Number 52 in Wissenschaftliche Berichte des Institutes für Fördertechnik und Logistiksysteme der Universität Karlsruhe (TH). Karlsruhe: Institut für Fördertechnik und Logistiksysteme der Universität Karlsruhe (TH).

Furmans, K. (2004). Zeitdiskrete Modellierung von Logistikprozessen. Technical report, Institut für Fördertechnik und Logistiksysteme, Universität Karlsruhe (TH).

Furmans, K., C. Huber and J. Wisser (2009). Queueing Models for manual order picking systems with blocking. *Logistics Journal September*(Online), p. 1–16.

Furmans, K., T. Stoll, F. Schönung and H. Hippenmeyer (2009). KARIS - dezentral gesteuert. Ein neuartiges Element für zukünftige Materialflusssysteme. *Hebezeuge Fördermittel 49*(6), p. 304–306.

Galka, S., A. Ulbrich and W. Günthner (2008). Performance Calculation for Order Picking Systems by Analytical Methods and Simulation. In: *Logistics and Supply Chain Management: Trends in Germany and Russia DR-LOG 08*, Moskow, Russia, p. 248–258. Publishing House of the St. Petersburg State Polytechnical University 2008.

Ghiani, G., G. Laporte and R. Musmanno (2004). *Introduction to Logistics Systems Planning and Control*. Chicester, West Sussex: WIley.

Gibson, D. R. and G. P. Sharp (1992). Order batching procedures. *European Journal of Operational Research 58*, p. 57–67.

Goetschalckx, M. and D. Ratliff (1990). Shared storage policies based the duration stay of unit loads. *Management Science 36*(9), p. 1120–1132.

Goetschalckx, M. and H. Ratliff (1991). Optimal lane depths for single and multiple products in block stacking storage systems. *IIE Transactions 23*(3), p. 245–258.

Goetschalckx, M. and H. D. Ratliff (1988). Order Picking In An Aisle. *IIE Transactions 20*(1), p. 53–62.

Goodman, L. A. (1960). On the Exact Variance of Products. *Journal of the American Statistical Association 55*(292), p. 708–713.

Gordon, W. J. and G. F. Newell (1967). Closed Queueing Systems with Exponential Servers. *Operations Research 15*(2), p. 254–265.

Gross, D. and C. M. Harris (1998). *Fundamentals of queueing theory* (3 ed.). John Wiley & Sons, Inc.

Gu, J., M. Goetschalckx and L. F. McGinnis (2007). Research on warehouse operation: A comprehensive review. *European Journal of Operational Research 177*, p. 1–21.

Gu, J., M. Goetschalckx and L. F. McGinnis (2010). Research on warehouse design and performance evaluation: A comprehensive review. *European Journal of Operational Research 203*, p. 539–549.

Gudehus, T. (1976). Staueffekte vor Transportknoten. *Zeitschrift für Operations Research 20*, p. B207–B252.

Gudehus, T. (2005). *Logistik: Grundlagen - Strategien - Anwendungen* (3 ed.). Berlin: Springer.

Gudehus, T. (2008). Kommissioniersysteme. In: D. Arnold, H. Isermann, A. Kuhn, H. Tempelmeier, and K. Furmans (eds.), *Handbuch Logistik* (3 ed.)., Chapter C2.3, p. 668–681. Berlin: Springer.

Gue, K. (2006). Very high density storage systems. *IIE Transactions 38*(1), p. 93–104.

Gue, K. (2010). Research - Very high density storage systems. http://web.mac.com/krgue.

Gue, K. and B. S. Kim (2007). Puzzle-Based Storage Systems. *Naval Research Logistics 54*(5), p. 556–567.

Gue, K. R. and R. D. Meller (2009). Aisle configurations for unit-load warehouses. *IIE Transactions 41*(3), p. 171–182.

Gue, K. R., R. D. Meller and J. D. Skufca (2006). The effects of pick density on order picking areas with narrow aisles. *IIE Transactions 38*(10), p. 859–868.

Hall, R. W. (1993). Distance approximations for routing manual pickers in a warehouse. *IIE Transactions 25*(4), p. 76–87.

Hillier, Frederick S. ; Lieberman, G. J. (1997). *Operations research : Einführung* (5 ed.). Internationale Standardlehrbücher der Wirtschafts- und Sozialwissenschaften. München: Oldenbourg.

Hoogendoorn, S. P. and P. Bovy (2001). State-of-the-art of vehicular traffic flow modelling. *Proceedings of the Institution of Mechanical Engineers, Part I: Journal of Systems and Control Engineering 215*(4), p. 283–303.

Hsieh, L.-f. and L. Tsai (2006). The optimum design of a warehouse system on order picking efficiency. *International Journal of Advanced Manufacturing Technology 28*, p. 626–637.

Hwang, H., Y. H. Oh and Y. K. Lee (2004). An evaluation of routing policies for order-picking operations in low-level picker-to-part system. *International Journal of Production Research 42*(18), p. 3873–3889.

Ivanović, G. (2007). Aisle design for unit-load warehouses with multiple pickup and deposit points. Master Thesis, Auburn University.

Jackson, J. R. (1957, August). Networks of Waiting Lines. *Operations Research 4*, p. 518–521.

Jane, C. C. (2000). Storage location assignment in a distribution center. *International Journal of Physical Distribution & Logistics 30*(1), p. 55–71.

Jane, C. C. and Y. W. Laih (2005). A clustering algorithm for item assignment in a synchronized zone order picking system. *European Journal of Operational Research 166*, p. 489–496.

Jarvis, J. M. and E. D. McDowell (1991). Optimal Product Layout in an Order Picking Warehouse. *IIE Transactions 23*(1), p. 93–102.

Johnson, N. L. and S. Kotz (1977). *Urn Models and Their Application - An Approach to Modern Discrete Probability Theory.* John Wiley & Sons, Inc.

Kallina, C. and J. Lynn (1976). Application of the Cube-per-Order Index Rule for Stock Location in a Distribution Warehouse. *Interfaces 7*(1), p. 37–46.

Kendall, D. G. (1953). Stochastic Processes Occurring in the Theory of Queues and their Analysis by the Method of the Imbedded Markov Chain. *The Annals of Mathematical Statistics 24*(3), p. 338–354.

Kleinrock, L. (1976). *Queueing Systems, Vol. 2: Computer Applications.* Wiley, New York.

Knott, K. and R. J. Sury (1987). A study of work-time distributions on unpaced tasks. *IIE Transactions 19*(3), p. 50–55.

Koo, P.-H. (2009). The use of bucket brigades in zone order picking systems. *OR Spectrum 21*, p. 759–774.

Krishnakumar, B. and C. J. Malmborg (1989). Modelling the service process in a multi-address warehousing system. *Applied Mathematical Modelling 13*(July), p. 386–396.

Kruskal, W. H. and W. A. Wallis (1952). Use of ranks in One-Criterion Variance Analysis. *Journal of the American Statistical Association 47*(260), p. 582–621.

Kunder, R. and T. Gudehus (1975). Mittlere Wegzeiten beim eindimensionalen Kommissionieren. *Zeitschrift für Operations Research 19*, p. B53–B72.

Le-Duc, T. (2005). *Design and Control of Efficient Order Picking Processes.* Ph.D. thesis, Erasmus Universiteit Rotterdam.

Le-Duc, T. and R. de Koster (2005a). Determining Number of Zones in a Pick-and-pack Orderpicking system. Technical report, Erasmus Research Institute of Management (ERIM) Report Series Research in Management.

Le-Duc, T. and R. de Koster (2005b). Travel distance estimation and storage zone optimization in a 2-block class-based storage strategy warehouse. *International Journal of Production Research 43*(17), p. 3561–3581.

Le-Duc, T. and R. de Koster (2005c). Travel Distance Estimation in Single-block ABC-Storage Strategy Warehouses. In: B. Fleischmann and A. Klose (eds.), *Distribution Logistics: Advanced Solutions to Practical Problems*, Lecture notes in economics and mathematical systems, p. 185–200. Berlin: Springer.

Le-Duc, T. and R. de Koster (2007). Travel time estimation and order batching in a 2-block warehouse. *European Journal of Operational Research 176*, p. 374–388.

Little, J. D. C. (1961). A Proof for the Queuing Formula: $L = \lambda \cdot W$. *Operations Research 9*(3), p. 383–387.

Liu, C.-M. (1999). Clustering techniques for stock location and order-picking in a distribution center. *Computers & Operations Research 26*, p. 989–1002.

Lüning, R. (2005). *Beitrag zur optmierten Gestaltung des Durchsatzes in Kommissioniersystemen für Stückgüter.* Göttingen: Cuvillier Verlag.

Malmborg, C. J. (1995). Optimization of cube-per-order index warehouse layouts with zoning contraints. *International Journal of Production Research 33*(2), p. 465–482.

Malmborg, C. J. and B. Krishnakumar (1990). A revised proof of optimaty for the cube-per-order index rule for stored item location. *Applied Mathematical Modelling 14*, p. 87–95.

Malmborg, C. J., B. Krishnakumar and G. R. Simons (1988). A mathematical overview of warehousing systems with single/dual order-picking cycles. *Applied Mathematical Modelling 12*, p. 2–8.

Marie, R. (1979). An Approximate Analytical Method for General Queueing Networks. *IEEE Transacions on Software Engineering 5*(5), p. 530–538.

Marie, R. (1980). Calculating equilibrium probabilities for $\lambda(n)|C_k|1|n$ queues. *ACM Computing Surveys 9*(2), p. 117–125.

Marie, R. A. and J. Pellaumail (1983). Steady-State Probabilities for a Queue with a General Service Distribution and State-Dependent Arrivals. *Software Engineering, IEEE Transactions on SE-9*(1), p. 109–113.

Marie, R. A., P. M. Snyder and W. J. Stewart (1982). Extensions and computational aspects of an iterative method. *SIGMETRICS Perform. Eval. Rev. 11*, p. 186–194.

Mayer, S. H. (2009). *Development of a completely decentralized control system for modular continuous conveyor systems.* Ph.D. thesis, Universität Karlsruhe, Karlsruhe.

Mellema, P. M. and C. A. Smith (1988). Simulation analysis of narrow-aisle order selection systems. In: M. Abrams, P. Haigh, and J. Comfort (eds.), *Proceedings of the 1988 Winter Simulation Conference*, p. 597–602.

Murrell, K. F. H. (1962). Operator variability and its industrial consequences. *International Journal for Production Research 1*(3), p. 39–55.

Nagel, K. and M. Schreckenberg (1992). A cellular automaton model for freeway traffic. *Journal de Physique 2*(12), p. 2221–2229.

Ohno, T. (1988). *Toyota production system: beyond large-scale production*. New York: Productivity Press.

Onvural, R. O. (1993). Queueing Networks with Finite Capacities. In: *Performance/SIGMETRICS Tutorials*, p. 411–434.

Onvural, R. R. (1990). Survey of closed queueing networks with blocking. *ACM Computing Surveys 22*(2), p. 83–121.

Ottjes, J. A. and E. Hoogenes (1988). Order Picking and Traffic Simulation in Distribution Centers. *International Journal of physical distribution & logistics management 18*(4), p. 14–21.

Pan, J. C.-H., J.-S. Chen and S.-Y. Lin (2005). Storage assignment in a narrow-aisle warehouse with multi-picker situation. *Journal of Industrial and Business Management 1*(1), p. 1–10.

Pan, J. C.-H. and P.-H. Shih (2008). Evaluation of the throughput of a multiple-picker order picking system with congestion consideration. *Computers & Industrial Engineering 55*(2), p. 379–389.

Pandit, R. and U. S. Palekar (1991). Response Time Considerations for Optimal Warehouse Layout Design. *Journal of Engineering for Industry 115*(August), p. 322–328.

Parikh, P. J. (2006). *Designing Order Picking Systems for Distribution Centers*. Ph.D. thesis, Virginia Polytechnic Institute and State University.

Parikh, P. J. and R. D. Meller (2008). Selecting between batch and zone order picking strategies in a distribution center. *Transportation Research Part E 44*, p. 696–719.

Parikh, P. J. and R. D. Meller (2009). Estimating picker blocking in wide-aisle order picking systems. *IIE Transactions 41*(3), p. 232–246.

Parikh, P. J. and R. D. Meller (2010). A note on worker blocking in narrow-aisle order picking systems when pick time is non-deterministic. *IIE Transactions 42*(6), p. 392 – 404.

Perros, H. G. (1989). A bibliography of papers on queueing networks with finite capacity queues. *Performance Evaluation 10*, p. 255–260.

Perros, H. G. (1994). *Queueing networks with blocking.* New York, NY, USA: Oxford University Press, Inc.

Petersen, C. G. (1997). An evaluation of order picking routeing policies. *International Journal of Operations & Production Management 17*(11), p. 1098–1111.

Petersen, C. G. (1999). The impact of routing and storage policies on warehouse efficiency. *International Journal of Operations & Production Management 19*(10), p. 1053–1064.

Petersen, C. G. (2002). Considerations in order picking zone configuration. *International Journal of Operations & Production Management 22*(7), p. 793–805.

Petersen, C. G. and G. Aase (2004). A comparison of picking, storage, and routing policies in manual order picking. *International Journal of Production Economics 92*(9), p. 11–19.

Petersen, C. G. and R. W. Schmenner (1999). An Evaluation of Routing and Volume-based Storage Policies in an Order Picking Operation. *Decision Sciences 30*(2), p. 481–501.

Rall, B. (1998). *Analyse und Dimensionierung von Materialflußsystemen mittels geschlossener Warteschlangennetze.* Ph.D. thesis, Universität Karlsruhe.

Ramin, S., B. R. Haverkort and P. Reinelt (2007). A Fixed-Point Algorithm for Closed Queueing Networks. In: *Formal Methods and Stochastic Models for Performance Evaluation,* Volume 4748 of *Lecture Notes in Computer Science,* p. 154–170. Springer.

Rana, K. (1990). Order Picking in Narrow-Aisle Warehouses. *International Journal of Physical Distribution & Logistics Management 20*(2), p. 9–15.

Ratliff, H. D. and A. S. Rosenthal (1983). Order-Picking in a Rectangular Warehouse: A Solvable Case of the Traveling Salesman Problem. *Operations Research 31*(3), p. 507–521.

Reiser, M. and S. Lavenberg (1980). Mean-Value Analysis of Closed Multichain Queuing Networks. *Journal of the ACM 27*(2), p. 313–322.

Roodbergen, K. J. (2001). *Layout and Routing Methods for Warehouses*. Ph.D. thesis, Erasmus Universiteit Rotterdam.

Roodbergen, K. J. and R. de Koster (2001a). Routing methods for warehouses with multiple cross aisles. *International Journal of Production Research 39*, p. 1865–1883.

Roodbergen, K. J. and R. de Koster (2001b). Routing order pickers in a warehouse with a middle aisle. *European Journal of Operational Research 133*, p. 32–43.

Roodbergen, K. J., G. P. Sharp and I. F. A. Vis (2008). Designing the layout structure of manual order picking areas in warehouses. *IIE Transactions 40*(11), p. 1032–1045.

Roodbergen, K. J. and I. F. A. Vis (2006). A model for warehouse layout. *IIE Transactions 38*, p. 799–811.

Rosenwein, M. B. (1994). An application of cluster analysis to the problem of locating items within a warehouse. *IIE Transactions 26*(1), p. 101–103.

Rosenwein, M. B. (1996). A comparison of heuristics for the problem of batching orders for warehouse selection. *International Journal of Production Research 34*(3), p. 657–664.

Rouwenhorst, B., B. Reuter, V. Stockrahm, G. van Houtum, R. Mantel and W. Zijm (2000). Warehouse design and control: Framework and literature review. *European Journal of Operational Research 122*, p. 515–533.

Ruben, R. A. and F. R. Jacobs (1999). Batch Construction Heuristics and Storage Assignment Strategies for Walk/Ride and Pick Systems. *Management Science 45*(4), p. 575–596.

Sadowsky, V. (2007). *Beitrag zur analytischen Leistungsermittlung von Kommissioniersystemen*. Ph.D. thesis, Universität Dortmund.

Schleyer, M. (2007). *Discrete Time Analysis of Batch Processes in Material Flow Systems*. Ph.D. thesis, Universität Karlsruhe.

Schlick, C., R. Bruder and H. Luczak (2010). *Arbeitswissenschaft* (3 ed.). Berlin Heildelberg: Springer.

Schmidt, L. C. and J. Jackman (2000). Modeling recirculating conveyors with blocking. *European Journal of Operational Research 124*, p. 422–436.

Schmidt, T., K. Turek, M. Krengel and M. Schmauder (2010). *Beschreibung der Dynamik manueller Operationen in logistischen Systemen*. Technical report, Institut für Technische Logistik und Arbeitssysteme, Fakultät Maschinenwesen, Technische Universität Dresden.

Schulte, C. (2009). *Logistik: Wege zur Optimierung der Supply Chain* (5 ed.). München: Vahlen.

Schulte, J. (1996). *Berechnungsgrundlagen konventioneller Kommissioniersysteme*. Ph.D. thesis, Universität Dortmund.

Sheth, V. S. (1995). *Facilities planning and materials handling: methods and requirements*. New York: Marcel Dekker.

Skufca, J. D. (2005). k Workers in a Circular Warehouse: A Random Walk on a Cricle, without passing. *SIAM Review 47*(2), p. 301–314.

Stadler, H. (1996). An operational planning concept for deep lane storage systems. *Production and Operations Management 5*(3), p. 266–282.

Suri, R. (1983). Robustness of Queueing Network Formulas. *Journal of the ACM 30*(3), p. 564–594.

Takahashi, Y., H. Miyahara and T. Hasegawa (1980). An Approximation Method for Open Restricted Queueing Networkd. *Operations Research 28*(3-Part-I), p. 594–602.

Tempelmeier, H. and H. Kuhn (1993). *Flexible manufacturing systems: decision support for design and operation*. John Wiley & Sons, Inc.

ten Hompel, M. and K. Hömberg (2008). Übersicht analytischer Berechnungsverfahren in Kommissioniersystemen. In: P. Nyhuis (eds.), *Beiträge zu einer Theorie der Logistik*, p. 391–408. Berlin: Springer.

ten Hompel, M., T. Schmidt and L. Nagel (2007). *Materialflusssysteme: Förder- und Lagertechnik* (3 ed.). Berlin: Springer.

Thayalan, P. (2008). Comparative study of item storage policies, vehicle routing strategies and warehouse layouts under congestion. Master Thesis, State University of New York at Buffalo.

Thomas, F. (2008). Lagersysteme. In: D. Arnold, H. Isermann, A. Kuhn, H. Tempelmeier, and K. Furmans (eds.), *Handbuch Logistik* (3 ed.)., Chapter C2.2.3, p. 648–657. Berlin: Springer.

Tompkins, J. A. (2003). *Facilities Planning* (3 ed.). Hoboken, NJ: Wiley.

Turek, K. and M. Krengel (2008). Modellierung der Dynamik manueller Operationen in logistischen Systemen. In: *4. Fachkolloquium der Wissenschaftlichen Gesellschaft für Technische Logistik e. V.*, Dresden.

van den Berg, J. (1996). *Planning and Control of Warehousing Systems*. Ph.D. thesis, Universiteit Twente.

van den Berg, J., G. Sharp, A. Gademann and Y. Pochet (1998). Forward-reserve allocation in a warehouse with unit-load replenishments. *European Journal of Operational Research 111*, p. 98–113.

van den Berg, J. P. (1999). A literature survey on planning and control of warehousing systems. *IIE Transactions 31*, p. 751–762.

van den Berg, J. P. and W. Zijm (1999). Models for warehouse management: Classification and examples. *International Journal of Production Economics 59*, p. 519–528.

Vaughan, T. S. and C. G. Petersen (1999). The effect of warehouse cross aisles on order picking efficiency. *International Journal of Production Research 37*(4), p. 881–897.

Vroblefski, M., R. Ramesh and S. Zionts (2000). General Open and Closed Queueing Networks with Blocking: A Unified Framework for Approximation. *INFORMS Journal on computing 12*(4), p. 299–316.

Vroblefski, M., R. Ramesh and S. Zionts (2005). A unified framework for approximation of general telecommunication networks. *European Journal of Operational Research 163*(2), p. 482 – 502.

Waldmann, K.-H. and U. M. Stocker (2004). *Stochastische Modelle: eine anwendungsorientierte Einführung*. EMILeA-stat - Medienreihe zur angewandten Statistik. Berlin: Springer.

WarehouseExcellence-Study (2011). Data from the ongoing study. Technical report, Institut für Fördertechnik und Logistiksysteme, Karlsruhe Institute of Technology.

Weber, R. R. and G. Weiss (1994). The cafeteria process - tandem queues with 0-1 dependent service times and the bowl shape phenomenon. *Operations Research 42*(5), p. 895–912.

White, J. A. (1972). Optimum design of warehouses having radial aisles. *AIIE Transactions 4*(4), p. 333–336.

Whitt, W. (1983). The Queueing Network Analyzer. *The Bell System Technical Journal 62*(9), p. 2779–2815.

Wisser, J. (2009). *Der Prozess Lagern und Kommissionieren im Rahmen des Distribution Center Reference Model (DCRM)*. Ph.D. thesis, Universität Karlsruhe.

Won, J. and S. Olafsson (2005). Joint order batching and order picking in warehouse operations. *International Journal of Production Research 43*(7), p. 1427–1442.

Wäscher, G. (2004). Order Picking: A Survey of Planning Problems and Methods. In: H. Dyckhoff, R. Lackes, and J. Reese (eds.), *Supply Chain Management and Reverse Logistics*, p. 323–347. Berlin: Springer.

Zhang, M., R. Batta and R. Nagi (2009). Modeling of Workflow Congestion and Optimization of Flow Routing in a Manufacturing/Warehouse Facility. *Management Science 55*(2), p. 267–280.

List of Figures

List of Tables

A Estimation of Parameters for a Cox-2 Model

To use a Cox-2 model for approximating distributions with mean service rate μ and variability $c_{ab}^2 \geq 0.5$ we have to determine the parameters μ_1, μ_2 and q_1 of the Cox-2-Model. μ_1 denotes the mean service rate in the first phase, μ_2 represents the mean service rate in the second phase and q_1 indicates the probability that the customer visits phase 2 after having finished service in phase 1.

Without explicitly giving a derivation, Marie (1980) suggested the following parameters:

$$\mu_1 = 2\mu \qquad \mu_2 = \frac{\mu}{c_{ab}^2} \qquad q_1 = \frac{1}{2c_{ab}^2}$$

In the following, we will present a detailed derivation of these parameters. In line with Augustin and Büscher (1982) as well as Furmans (2000), we formulate three conditions that have to be fulfilled:

- The mean service time of the overall system has to equal the sum of the mean service times in the respective phases

$$\frac{1}{\mu} \overset{!}{=} \frac{1}{\mu_1} + \frac{q1}{\mu_2}$$

- The service times of the two phases have to be equal

$$\frac{1}{\mu_1} \overset{!}{=} \frac{q1}{\mu_2}$$

- The second moment of the overall service time has to equal the second moment of the Cox-2-Distribution.

$$\frac{1}{\mu^2}(1 + c_{ab}^2) \overset{!}{=} E(C^2)$$

A.1 Second Moment for a Cox-n-Distribution

We want to derive the second moment and estimate the parameters μ_1, μ_2 and q_1. Augustin and Büscher (1982) first transform the standard Cox-n-Representation into an equivalent representation. In this model we have n parallel paths and choose exactly one path i with probability p_i. Path 1 consists of one exponential server, path 2 consists of two serial exponential servers, ... and path n consists of n serial exponential servers. Let Z_i be an exponentially distributed random variable representing the service time in phase i. Let C be a random variable representing the service time of the overall Cox-n-Model.

Let $(1 - q_i)$ be the probability to exit the system after completing service in phase i. Then p_i indicates the probability to enter a specific path i and is given by:

$$p_i = (1 - q_i) \prod_{k=1}^{i-1} q_k$$

With these definitions the second moment of a Cox-n-Distribution is generally given by:

$$E(C^2) = \sum_{r=1}^{n} p_r \cdot E\left[\left(\sum_{i=1}^{r} Z_i\right)^2 \right]$$

$$E(C^2) = \sum_{i=1}^{n} p_r \cdot \left[\sum_{i=1}^{r} E(Z_i^2) + 2 \cdot \sum_{i=2}^{r} \sum_{j=1}^{i-1} E(Z_i \cdot Z_j) \right] \qquad \text{(A.1)}$$

A.2 Second Moment for a Cox-2-Distribution

For the special case of a Cox-2-Distribution, equation (A.1) simplifies to:

$$E(C^2) = p_1 \cdot E(Z_1^2) + p_2 \cdot \left[E(Z_1^2) + E(Z_2^2) + 2 \cdot E(Z_1 Z_2) \right] \qquad \text{(A.2)}$$

For two phases we can define the parameters p_1 and p_2 as follows:

$$p_1 = (1 - q_1) \qquad p_2 = q_1$$

and formulate the second moment as:

$$E(C^2) = (1 - q_1) \cdot E(Z_1^2) + q_1 \cdot \left[E(Z_1^2) + E(Z_2^2) + 2 \cdot E(Z_1 Z_2) \right] \quad (A.3)$$

In order to better define equation (A.3) we now have to consider the terms $E(Z_1^2)$, $E(Z_2^2)$ and $E(Z_1 Z_2)$ in more detail. As the phases are independent, i.e. all Z_i are independent, we can re-write the term $E(Z_1^2)$ as follows:

$$E(Z_1^2) = Var(Z_i) + E(Z_i)^2$$

Based on independence we can also set:

$$E(Z_1 Z_2) = E(Z_1) \cdot E(Z_2)$$

For an exponentially distributed random variable Z_i the variance and mean service time are defined by:

$$Var(Z_i) = \frac{1}{\mu_i^2} \qquad E(Z_i) = \frac{1}{\mu_i}$$

The second moment of the random variable Z_i can then be formulated as:

$$E(Z_1^2) = Var(Z_i) + E(Z_i)^2 = \frac{1}{\mu_i^2} + \left(\frac{1}{\mu_i} \right)^2 = \frac{2}{\mu_i^2} \qquad (A.4)$$

We can now re-formulate equation (A.3) as follows:

$$
\begin{aligned}
E(C^2) &= (1 - q_1) \cdot \left(\frac{2}{\mu_1^2} \right) + q_1 \cdot \left(\frac{2}{\mu_1^2} + \frac{2}{\mu_1 \mu_2} + \frac{2}{\mu_2^2} \right) \\
&= \frac{2}{\mu_1^2} + q_1 \cdot \left(\frac{2}{\mu_1 \mu_2} + \frac{2}{\mu_2^2} \right) \\
&= 2 \cdot \left(\frac{1}{\mu_1^2} + q_1 \cdot \left(\frac{1}{\mu_1 \mu_2} + \frac{1}{\mu_2^2} \right) \right) \qquad (A.5)
\end{aligned}
$$

Equation (A.5) is identical to the formulation given in the original literature (Augustin and Büscher 1982, p. 26). Based on the second moment $E(C^2)$, we can now use the three conditions formulated above to derive the parameters for the Cox-2-Distribution.

A.3 Derivation of the Parameters μ_1, μ_2 and q_1

We start with condition 2 (service time of the phases have to be equal) to derive an intermediate result for parameter q_1:

$$\frac{1}{\mu_1} \overset{!}{=} \frac{q1}{\mu_2}$$

$$q_1 = \frac{\mu_2}{\mu_1} \tag{A.6}$$

We use this result in condition 1 (mean service time of the overall system has to equal the sum of the mean service times in the respective phases) to get an intermediate result for μ_1:

$$\frac{1}{\mu} \overset{!}{=} \frac{1}{\mu_1} + \frac{q1}{\mu_2}$$

$$\Rightarrow \mu_1 = \frac{\mu\mu_2}{\mu_2 - \frac{\mu_2}{\mu_1}\mu} \tag{A.7}$$

We again use condition 1 to further characterize parameter q_1:

$$q_1 = \frac{\mu_2}{\frac{\mu\mu_2}{\mu_2 - \frac{\mu_2}{\mu_1}\mu}}$$

$$q_1 = \frac{\mu_2 \cdot \left(\mu_2 - \frac{\mu_2\mu}{\mu_1}\right)}{\mu\mu_2}$$

$$q_1 = \frac{\mu_2 - \frac{\mu_2}{\mu_1}\mu}{\mu}$$

$$q_1 = \frac{\mu_2 - q_1\mu}{\mu}$$

$$q_1 = \frac{\mu_2}{\mu} - q_1$$

$$2q_1 = \frac{\mu_2}{\mu}$$

$$q_1 = \frac{\mu_2}{2\mu} \tag{A.8}$$

Again we use condition 1 and get the final result for parameter μ_1:

$$\frac{1}{\mu} \overset{!}{=} \frac{1}{\mu_1} + \frac{q1}{\mu_2}$$

$$\frac{1}{\mu} \overset{!}{=} \frac{1}{\mu_1} + \frac{\frac{\mu_2}{2\mu}}{\mu_2}$$

$$\Rightarrow \mu_1 = 2\mu \tag{A.9}$$

Now we will use the final result from equation (A.9), the intermediate result from equation (A.8) and the formula for the second moment from equation (A.5) in condition 3 (the second moment of the overall service time has to equal the second moment of the Cox-2-Distribution):

$$\frac{1}{\mu^2}(1 + c_{ab}^2) \;\stackrel{!}{=}\; E(C^2)$$

$$\frac{1}{\mu^2}(1 + c_{ab}^2) \;=\; 2 \cdot \left(\frac{1}{\mu_1^2} + q_1 \cdot \left(\frac{1}{\mu_1\mu_2} + \frac{1}{\mu_2^2}\right)\right)$$

$$\frac{1}{\mu^2}(1 + c_{ab}^2) \;=\; 2 \cdot \left(\frac{1}{\mu_1^2} + q_1 \cdot \left(\frac{\mu_1 + \mu_2}{\mu_1\mu_2^2}\right)\right)$$

$$\frac{1}{\mu^2}(1 + c_{ab}^2) \;=\; 2 \cdot \left(\frac{1}{4\mu^2} + \frac{\mu_2}{2\mu} \cdot \left(\frac{2\mu + \mu_2}{2\mu\mu_2^2}\right)\right)$$

$$\frac{1}{\mu^2}(1 + c_{ab}^2) \;=\; 2 \cdot \left(\frac{1}{4\mu^2} + \frac{\mu_2}{2\mu} \cdot \left(\frac{1}{\mu_2^2} + \frac{1}{2\mu\mu_2}\right)\right)$$

$$\frac{1}{\mu^2}(1 + c_{ab}^2) \;=\; \frac{1}{2\mu^2} + 2\frac{\mu_2}{2\mu}\frac{1}{\mu_2^2} + 2\frac{\mu_2}{2\mu}\frac{1}{2\mu\mu_2}$$

$$\frac{1}{\mu^2}(1 + c_{ab}^2) \;=\; \frac{1}{2\mu^2} + \frac{1}{\mu\mu_2} + \frac{1}{2\mu^2}$$

$$\frac{1}{\mu^2}(1 + c_{ab}^2) - \frac{1}{2\mu^2} - \frac{1}{2\mu^2} \;=\; \frac{1}{\mu\mu_2}$$

$$\frac{1}{\mu^2}(1 + c_{ab}^2) - \frac{1}{\mu^2} \;=\; \frac{1}{\mu\mu_2}$$

$$\frac{1}{\mu^2}(1 + c_{ab}^2 - 1) \;=\; \frac{1}{\mu\mu_2}$$

$$\frac{\mu^2}{c_{ab}^2} \;=\; \mu\mu_2$$

$$\Rightarrow \mu_2 \;=\; \frac{\mu}{c_{ab}^2} \tag{A.10}$$

Finally we use the result of equation (A.10) in the intermediate result of equation (A.8) to get the final result for q_1:

$$q_1 \;=\; \frac{\mu_2}{2\mu}$$

$$q_1 \;=\; \frac{\frac{\mu}{c_{ab}^2}}{2\mu}$$

$$\Rightarrow q_1 \;=\; \frac{1}{2c_{ab}^2} \tag{A.11}$$

Observing equations (A.9), (A.10) and (A.11), we can see that these parameters equal those originally proposed by Marie (1980, p. 128). Experimental tests have shown that the parameters

$$\mu_1 = 2\mu \qquad \mu_2 = \frac{\mu}{c_{ab}^2} \qquad q_1 = \frac{1}{2c_{ab}^2}$$

produce good solutions concerning the approximation of both μ and c_{ab}^2.

B Fit-Functions for the Correction Factor f_{ZB}

We will briefly present the fit-functions used in the surface-fitting operation to find a functional relation between f_{ZB} and the input parameters ω as well as one of the parameters G^* and $c^2_{R,ZB}$.

Cosine2D(ω, G^*)

$$f_{ZB}^{true} \approx f_{ZB} = z_0 + A1 \cdot cos(\omega) + B1 \cdot cos(G^*) + A2 \cdot cos(2\omega)$$
$$+ C1 \cdot cos(\omega) \cdot cos(G^*) + B2 \cdot cos(2G^*) \quad \text{(B.1)}$$

Cosine2D(ω, $c^2_{R,ZB}$)

$$f_{ZB}^{true} \approx f_{ZB} = z_0 + A1 \cdot cos(\omega) + B1 \cdot cos(c^2_{R,ZB}) + A2 \cdot cos(2\omega)$$
$$+ C1 \cdot cos(\omega) \cdot cos(c^2_{R,ZB}) + B2 \cdot cos(2c^2_{R,ZB}) \quad \text{(B.2)}$$

Plane(ω, G^*)

$$f_{ZB}^{true} \approx f_{ZB} = z_0 + a \cdot \omega + b \cdot G^* \quad \text{(B.3)}$$

Plane(ω, $c_{R,ZB}^2$)

$$f_{ZB}^{true} \approx f_{ZB} = z_0 + a \cdot \omega + b \cdot c_{R,ZB}^2 \tag{B.4}$$

Power2D(ω, G^*)

$$f_{ZB}^{true} \approx f_{ZB} = z_0 + B \cdot \omega^C + D \cdot (G^*)^E + F \cdot \omega^C \cdot (G^*)^E \tag{B.5}$$

Power2D(ω, $c_{R,ZB}^2$)

$$f_{ZB}^{true} \approx f_{ZB} = z_0 + B \cdot \omega^C + D \cdot (c_{R,ZB}^2)^E + F \cdot \omega^C \cdot (c_{R,ZB}^2)^E \tag{B.6}$$

Gauss2D(ω, G^*)

$$f_{ZB}^{true} \approx f_{ZB} = z_0 + A \cdot e^{\left[-0.5 \cdot \left(\frac{\omega - x_c}{w_1}\right)^2 - 0.5 \cdot \left(\frac{G^* - y_c}{w_2}\right)^2\right]} \tag{B.7}$$

Gauss2D(ω, $c_{R,ZB}^2$)

$$f_{ZB}^{true} \approx f_{ZB} = z_0 + A \cdot e^{\left[-0.5 \cdot \left(\frac{\omega - x_c}{w_1}\right)^2 - 0.5 \cdot \left(\frac{c_{R,ZB}^2 - y_c}{w_2}\right)^2\right]} \tag{B.8}$$

Poly2D(ω, G^*)

$$\begin{aligned} f_{ZB}^{true} \approx f_{ZB} = {} & z_0 + A \cdot \omega + B \cdot G^* \\ & + C \cdot \omega^2 + D \cdot (G^*)^2 + F \cdot \omega \cdot G^* \end{aligned} \tag{B.9}$$

$$f_{ZB}^{true} \approx f_{ZB} = z_0 + A \cdot \omega + B \cdot c_{R,ZB}^2$$
$$+C \cdot \omega^2 + D \cdot (c_{R,ZB}^2)^2 + F \cdot \omega \cdot c_{R,ZB}^2 \qquad \text{(B.10)}$$